Biochemical Mechanisms of Aluminium Induced Neurological Disorders

Edited by

Touqeer Ahmed

Department of Healthcare Biotechnology
Atta-ur-Rahman School of Applied Biosciences
National University of Sciences and Technology
Islamabad, Pakistan

Biochemical Mechanisms of Aluminium Induced Neurological Disorders

Editor: Touqeer Ahmed

ISBN (Online): 978-1-68108-883-9

ISBN (Print): 978-1-68108-884-6

ISBN (Paperback): 978-1-68108-885-3

©2021, Bentham Books imprint.

Published by Bentham Science Publishers – Sharjah, UAE. All Rights Reserved.

need for a court order if at any point you breach any terms of this License Agreement. In no event will any delay or failure by Bentham Science Publishers in enforcing your compliance with this License Agreement constitute a waiver of any of its rights.

3. You acknowledge that you have read this License Agreement, and agree to be bound by its terms and conditions. To the extent that any other terms and conditions presented on any website of Bentham Science Publishers conflict with, or are inconsistent with, the terms and conditions set out in this License Agreement, you acknowledge that the terms and conditions set out in this License Agreement shall prevail.

Bentham Science Publishers Ltd.
Executive Suite Y - 2
PO Box 7917, Saif Zone
Sharjah, U.A.E.
Email: subscriptions@benthamscience.net

BENTHAM SCIENCE

CONTENTS

PREFACE

This book "Biochemical Mechanisms of Aluminium Induced Neurological Disorders" is composed of six chapters contributed by reputed scientists working in the field of metals neurotoxicity and studying the role of metals on various neurological disorders. The salient features of this book which make it unique are– it highlights the basic as well as clinical mechanisms of metals induced neurotoxicity in various neurological disorders, and another unique feature of this book is that it discusses the role of Aluminium induced neurological disorders, alone and in combination with other toxic metals as well as with the high fat diet intake, thus adding diversity and unraveling new features of metals neurotoxicity in real scenarios.

Aluminium is present in the earth's crust and it is a well known environmental toxin/metal which has been found to be associated with various neurological disorders. Aluminium has been found to be a very strong risk factor for the development of Alzheimer's disease. It is known to cause neurotoxicity by various mechanisms, which are highlighted in this book. Cholinergic system impairment seems to be prominent; however, other mechanisms and pathways are also discussed and elaborated in this book. This book also covers missing aspects of the blood brain barrier and developmental toxicity, which are not very well studied areas related to Aluminium exposure alone or in combination with other metals. Knowing the pharmacokinetics of Aluminium and other metals are important aspects that can help us to understand the exposure of metals and their brain entry mechanisms, thus, opening up new horizons for the development of therapeutic options. Finally, I would say that this comprehensive book is well balanced and well written in developing understandings of basic and clinical mechanisms of Aluminium induced neurological disorders.

Touqeer Ahmed
Department of Healthcare Biotechnology
Atta-ur-Rahman School of Applied Biosciences
National University of Sciences and Technology
Islamabad, Pakistan

List of Contributors

Abida Zulfiqar	Neurobiology Laboratory, Department of Healthcare Biotechnology, Atta-ur-Rahman School of Applied Biosciences, National University of Sciences and Technology, Sector H-12, Islamabad 44000, Pakistan
Amna Liaqat	Neurobiology Laboratory, Department of Healthcare Biotechnology, Atta-ur-Rahman School of Applied Biosciences, National University of Sciences and Technology, Sector H-12, Islamabad 44000, Pakistan
Armeen Hameed	Neurobiology Laboratory, Department of Healthcare Biotechnology, Atta-ur-Rahman School of Applied Biosciences, National University of Sciences and Technology, Sector H-12, Islamabad 44000, Pakistan
Fatima Javed Mirza	Neurobiology Laboratory, Department of Healthcare Biotechnology, Atta-ur-Rahman School of Applied Biosciences, National University of Sciences and Technology, Sector H-12, Islamabad 44000, Pakistan
Ghazal Iqbal	Neurobiology Laboratory, Department of Healthcare Biotechnology, Atta-ur-Rahman School of Applied Biosciences, National University of Sciences and Technology, Sector H-12, Islamabad 44000, Pakistan
Laraib Liaquat	Multidisciplinary Research Lab, Bahria University Medical and Dental College, Bahria University, Karachi, Pakistan
Rida Nisar	HEJ Research Institute of Chemistry, International Center for Chemical and Biological Sciences, University of Karachi, Karachi, Pakistan
Saadia Zahid	Neurobiology Laboratory, Department of Healthcare Biotechnology, Atta-ur-Rahman School of Applied Biosciences, National University of Sciences and Technology, Sector H-12, Islamabad 44000, Pakistan
Saida Haider	Neurochemistry and Biochemical Neuropharmacology Research Unit, Department of Biochemistry, University of Karachi, Karachi, Pakistan
Sanila Amber	Neurobiology Laboratory, Department of Healthcare Biotechnology, Atta-ur-Rahman School of Applied Biosciences, National University of Sciences and Technology, Sector H-12, Islamabad 44000, Pakistan
Sara Ishaq	Neurobiology Laboratory, Department of Healthcare Biotechnology, Atta-ur-Rahman School of Applied Biosciences, National University of Sciences and Technology, Sector H-12, Islamabad 44000, Pakistan
Syeda Mehpara Farhat	Department of Biological Sciences, National University of Medical Sciences, Rawalpindi-46000, Pakistan
Touqeer Ahmed	Neurobiology Laboratory, Department of Healthcare Biotechnology, Atta-ur-Rahman School of Applied Biosciences, National University of Sciences and Technology, Sector H-12, Islamabad 44000, Pakistan
Tuba Sharf Batool	Atta-ur-Rahman School of Applied Biosciences, National University of Sciences and Technology, Islamabad, Pakistan
Zehra Batool	Dr. Panjwani Center for Molecular Medicine and Drug Research, International Center for Chemical and Biological Sciences, University of Karachi, Karachi, Pakistan

Biochemical Mechanisms of Aluminium and Other Metals Exposure, Their Brain Entry Mechanisms, Effects on Blood Brain Barrier and Important Pharmacokinetic Parameters in Neurological Disorders

Sara Ishaq[1]**, Amna Liaqat**[1]**, Armeen Hameed**[1] **and Touqeer Ahmed**[1,*]

[1] *Neurobiology Laboratory, Department of Healthcare Biotechnology, Atta-ur-Rahman School of Applied Biosciences, National University of Sciences and Technology, Sector H-12, Islamabad 44000, Pakistan*

Abstract: Evolution of life has resulted in a strong association between environmental metals and the biological processes taking place in the human body. Some of these metals are essential for the survival of human life, while many others can pose harmful effects on the body if exposed continuously. These toxic metals include Aluminium (Al), Arsenic (As), Lead (Pb), Mercury (Hg), Cadmium (Cd) etc. Upon entry into the brain, these metals lead to the development of many neurological disorders by increasing the levels of ROS, disturbing calcium ion efflux, causing mitochondrial dysfunction and activating an immunogenic response. These metals also cause a decrease in the levels of certain antioxidants in the brain like glutathione, superoxide dismutase and catalase. Moreover, the decrease in the level of certain genes like brain derived neurotropic factor (BDNF) due to metals neurotoxicity can also cause depletion of the memory and other cognitive functions leading to many neurodegenerative diseases like Alzheimer's disease (AD), Parkinson's disease (PD), etc. The following chapter explains the pharmacokinetic mechanisms involved in metals induced neurotoxicity leading to different neurological disorders.

Keywords: Neurodegeneration, Metals Accumulation, Metals Toxicity, Metals Pharmacokinetics, Metals Distribution.

---

* **Correspondence author Touqeer Ahmed PhD**: Neurobiology Laboratory, Department of Healthcare Biotechnology, Atta-ur-Rahman School of Applied Biosciences, National University of Sciences and Technology, Sector H-12, Islamabad- 44000, Pakistan; Tel: +92-51-9085-6141, Fax: +92-51-9085-6102; E-mail: touqeer.ahmed@asab.nust.edu.pk

INTRODUCTION

Metals and their Evolution in Biological Processes

Metals have been associated with biological systems for billions of years and this association has also been known to evolve with time. Many life processes include a variety of naturally occurring metal complexes in different ways [1]. Major metals like iron, zinc, magnesium, manganese etc. and minor metal ions like copper, nickel, cobalt, molybdenum, tungsten, etc., have been incorporated into the living organisms by the interplay of their metabolic pathways with the products of biogeochemical weathering [2]. Organisms are now able to adapt or die due to the natural development of these metals and other chemicals. Many important life processes of current organisms especially, the metabolic processes require redox reactions which are dependent on the presence of these metals as they have a tendency to lose or gain electrons [3].

Metals are so central in the cellular processes that almost 30% of the overall body proteins are metallo-proteins. Almost 40% of all enzymatic reactions require metals and at least one step of all the biological pathways involve a metal [4]. For example, calcium is not only required for strong bones and teeth but is also involved in reducing muscle cramps and triggering a number of cellular processes. Similarly, many of the cellular activities are dependent on magnesium which is the most abundant element inside the cells after potassium. The biological processes taking place in the nucleus involve metals like calcium, magnesium, copper, zinc, iron and manganese which are present there, in detectable amounts, *i.e.*, 10^{-2}-10^{-4}mol. These metals bind to the DNA and RNA in the cells, even RNA's active configuration is also dependent upon the concentrations of magnesium and manganese [5]. Magnesium is also responsible for providing energy to millions of cells in the animal and plant bodies by the activation of the production of ATP. It is also involved in some other processes like the process of DNA polymer synthesis along with other divalent metal ions like zinc and manganese [6]. Some of the important functions of all of these metals are given in Table **1** in detail. Thus, the metals are considered to be essential for the biological system as without them the system may collapse.

Table 1. Some of the important functions of essential metals and their related deficiency problems inside the body.

Metals		Important Biological Functions	Deficiency Issues	References
Major Essential Metals	**Calcium (Ca2+)**	• Provides strength to bones and teeth • Involved as a second messenger in signal transduction pathways like neurotransmitter release, muscle contraction, fertilization, hormonal release etc. • Acts as enzymes cofactor • Involved in blood coagulation process • Maintains potential difference across excitable cell membranes	• Seizures • Depression • Dental Problems • Osteopenia and Osteoporosis • Various skin conditions • Painful premenstrual syndrome • Chronic joint and muscle pain • Bones weaknesses and fractures • Muscular disability	[7-10]
	Sodium (Na+)	• Maintains blood, plasma and other body fluids' homeostasis • Involved in signal transduction of the central nervous system by controlling renin- angiotensin system and atrial natriuretic peptide • Involved in the transport of solutes across cell membranes via Na2+/K+ pump • Involved in body's buffer system via Na2+/K+ pump	• Headache • Confusion • Seizures • Nausea and vomiting • Muscles weakness, spasms or cramps • Muscular irritability and restlessness • Loss of energy, drowsiness and fatigue • Coma	[11-13]
	Potassium (K+)	• Involved in electrolyte metabolism along with sodium and chloride ions • Help in conduction of nerve impulses • Acts as a cofactor of enzymes • Controls the transport of essential elements via Na2+/K+ pumps • Involved in electrical signaling via potassium channels	• Muscle paralysis • Cardiac arrhythmias • Mood disorders • Tingling and numbness in hands and feet • Breathing difficulties • Weakness and fatigue • Muscles cramps and spasms • Digestive problems	[12, 14, 15]

	Magnesium (Mg2+)	• Involved in muscle contraction, neuromuscular conduction, glycemic control and myocardial contraction • Maintains the blood pressure • Acts as a cofactor or more than 300 enzymes • Involved in energy production (ATP) • Acts as active trans- membrane transporter for other ions • Involved in synthesis of nuclear material • Involved in bone and formation	• Muscle tremors, spasms and cramps • Nausea and vomiting • Weakness and fatigue • Mood disorders • Seizures • Depression • Encephalopathy • Agitation • Numbness and tingling in hands and feet • ECG loss and cardiac arrhythmias • Loss of appetite • Hypokalemia and hypocalcaemia	[16-19]

Minor Essential Metals or Trace Metals	Iron (Fe2+)	• Main constituent (80%) of oxygen carrying protein (hemoglobin) in the blood • Involved in hundreds of enzymatic reactions like oxygen transport, DNA synthesis and electron transport	• Iron deficiency anemia • Fatigue • Hair loss • Dizziness • Twitches • Brittle or grooved nails • Pagophagia • Impaired immune system • Restless legs syndrome	[15, 20, 21]
	Manganese (Mn2+)	• Essential metal for a number of intracellular activities • Acts as cofactor in many enzymatic reactions including metabolism, regulation of cellular energy, reproduction and growth of bones and connective tissues as glutamine synthetase, manganese superoxide dismutase and arginase • Manganese superoxide dismutase protects mitochondria from toxic oxidants • Glutamine synthetase is the most abundant manganese enzyme in the body which is involved in brain functions	• Impaired growth • Impaired reproductive functions • Skeletal abnormalities • Impaired glucose tolerance • Altered carbohydrate and lipid metabolism • Decreased serum cholesterol levels • Transient skin	[22-25]
	Zinc (Zn2+)	• Takes part in enzymatic reactions of more than 300 proteins like superoxide dismutase, carbonic anhydrase, alkaline phosphatase etc. • Plays catalytic role in acid- base reactions in association with RNA-polymerase • Involved in organizing tertiary structures of proteins and regulation of gene expression via zinc fingers • Major regulatory ion among the redox inert metal ions Na+, K+, Mg2+, Ca2+, sharing its signaling capacity with calcium. • Have the ability to act as second messengers thus help calcium ions in signaling pathways	• Hypogonadism • Dwarfism • Decreased resistance against diseases • Eye and skin lesions • Diarrhea • Hair loss • Delayed sexual maturation • Impotence	[26-28]

	Copper (Cu2+)	• Important in various enzymatic reactions, particularly as an electron donor • It is specifically associated with electron transporter enzyme cytochrome c in respiratory chain • Involved in the formation of connective tissue such as collagen and keratin • Involved in vitamin B12 rearrangements • Carries out reduction reactions	• Anemia (unresponsive to iron therapy) • Neutropenia • Abnormality in the process of cell renewal • Osteoporosis and other bone diseases • Impaired growth • Neurological diseases • Loss of pigmentation	[29-34]

Metals Induced Neurotoxicity

One quarter of 20 top health conditions around the world are neurological diseases. About one million people covering almost 1% of global prevalence are suffering from these diseases. The main cause is poor hygienic conditions. Metals are one of the most common sources of environmental contamination. These metals affect the brain especially in children, mostly by the production of Reactive Oxygen Species (ROS) or by damaging the DNA and proteins structures [55]. Al is the most common neurotoxin leading to neurodegeneration, cognitive dysfunction, Blood Brain Barrier (BBB) disruption, neuroinflammation, impaired cholinergic projections and neuronal death [56 - 59] as is among the top toxicants and is involved in many neuropathies. It causes demyelination of axons, encephalopathy, cognitive impairments, irritability and headaches [60, 61]. Pb toxicity is another major concern, especially in developed countries, due to its non-biodegradable nature. Its high levels lead to decreased IQ, muscular dysfunctions and irritability, convulsions, hallucinations, dull personality, ataxia, headaches, coma memory loss, etc [62 - 65]. Hg has also been reported to be involved in many neurological diseases especially in the form of methylmercury interacting with ROS production and release in the brain. Hg has been found to cause fatigue, irritability, tremors, headaches, cognitive dysfunction and loss of hearing, hallucinations, dysarthria and even death [66, 67]. Cd has also been found to be involved in major neurological symptoms. It is found to be associated with hallucinations, headaches, vertigo, Parkinson's like symptoms, slow vasomotor functions, muscular and learning disabilities [54, 68].

ABSORPTION OF TOXIC METALS IN THE BRAIN

Metals are bliss as well as harmful to the brain. In view of both harmful and nurturing aspects of the brain, there is the development of a variety of protective mechanisms to check the uptake of metals from blood as well as its distribution within brain tissues. The reason for such tight regulation is that brain can regenerate its cellular components up to a limit and metals toxicity can cause irreversible damage to neurons [69]. Metals levels and homeostasis in the brain is maintained to a specific limit by a structural barrier which under normal conditions prevent any infiltration of unwanted substances from blood to cross the barrier and enter the brain. Structurally these barriers are of two kinds: BBB and blood cerebrospinal fluid barrier (BCB). Former is the barrier between systemic circulation and interstitial fluid while later is between systemic circulation and cerebrospinal fluid [70].

Transport of Toxic Metals Across Blood Brain Barrier

BBB serves as an essential barrier serving both metabolic and physical roles in maintaining the normalized function of central nervous system (CNS). Main targets of this barrier involve restraining the paracellular movement of toxic metals and other hydrophilic molecules from blood [71]. Integrity of the BBB is maintained by certain structural elements such as cerebral endothelial cells (CECs), (Fig. 1).pericytes and glial end-foot [72]. Among these elements, CECs are important, especially the tight junctions (TJs) present in the CECs. These are particularly involved in maintaining vascular permeability. At the molecular level, protein components of TJs include claudin, occludin, and junctional adhesion molecules, and cytosolic proteins. All these serve as transmembrane proteins. Proteins from cellular compartments of neurons interact with these transmembrane proteins and form multi-protein complexes, which in turn are linked to the actin polymers and its associated actin binding protein, collectively called actin cytoskeleton [73].

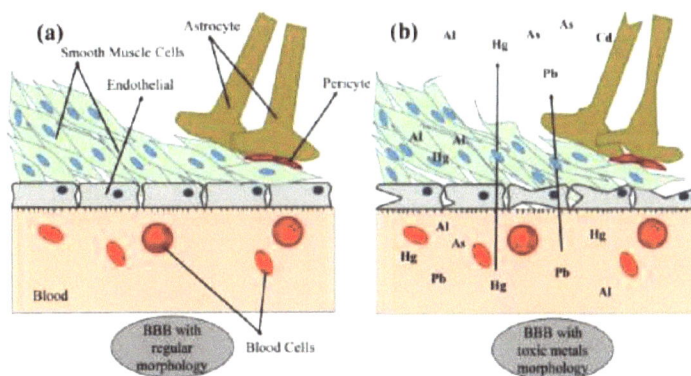

Fig. (1). Blood brain barrier shown as under (a) normal physiological conditions and (b) under conditions of metals induced neurotoxicity.

Toxic metals can target certain crucial regions of brain by gaining access by imitating the effect of BBB. Generally, toxic metals can be absorbed by the GI tract or through the lungs and then moving to blood circulation. Once into the blood circulation, gaining access to brain regions is entirely dependent upon the structural integrity of BBB. If this barrier is compromised, then metals can get into the choroid plexuses and cerebrospinal fluid [74]. Although there are specific mechanisms to check the flux of metals and other nutrients in and out of the brain and especially of toxic chemicals [75]. Toxic metals can cause poisoning of BBB and resulting in cerebral hemorrhage, vascular damage and, importantly, the destruction of endothelial TJs (Fig. **1**). It leads to excessive leakage of metals from the blood into the brain. Most toxic metals are also reported to accumulate in the brain, especially Al, Cd, Hg and As can easily accumulate in the brain at higher concentrations [76]. Another strategy adopted by metals to gain access to brain is by mimicking the behavior of other essential metals or other nutrients and then utilizing the ionic transporters [77].

Transport of Toxic Metals Across Blood Cerebrospinal Fluid Barrier

The primary element of the blood cerebrospinal fluid barrier is the choroid plexus (CP). Physiologically CP is a dense network of capillaries together with ependymal cells and is located in the cerebral ventricles. CP is both the element of BCB and a producer of cerebrospinal fluid. External side of BCB faces the systemic circulation while the inner side is in contact with the cerebral sections. However, the barrier keeps both sides completely out-of-the-way from each other [78]. Hence structural integrity of BCB is also vital for maintaining a normal homeostasis level in brain [79]. Chemical constituents and metals levels are strictly regulated within the brain by balancing the movement of materials across the barrier from blood into the CSF and vice versa. It acts in a bidirectional way and transport substances. It has been studied that if there are any impairments in barrier structure, it leads to leakage of metals and can cause clinical encephalopathies [80]. Metals are also known to accumulate in the CP and then can find their way to other parts of brain, especially the cerebral parts and hippocampus [80].

MAPPING BRAIN REGIONS WITH TOXIC METALS DISTRIBUTION

Brain is the vital organ processing essential functions such as learning, memory formation, movement and other processes. Environmental pollutants and toxins can cause brain damage resulting in compromised brain functions. Toxic metals such as Pb, Cd, As, Al and Hg have shown to be neurotoxic when exposed to higher concentrations. Metals neurotoxic effects are demonstrated when these toxic metals cross BBB and BCB through different mechanisms; moreover, these metals have the tendency to accumulate in the brain. This accumulation in fact, is more lethal to brain. Different toxic metals get accumulated in brain at different concentrations, as indicated in many scientific studies. Even considering the sub cellular organelles of neurons, metals concentrations are variedly accumulated.

Aluminium

Al shows up to be highly accumulated in hippocampal areas and corpus striatum and is then followed by other regions in the order of decreasing concentration as brain stem > cerebral cortex > cerebellum [81]. Going one step further and talking about sub cellular structures, then Al even shows different levels in different organelles. Highest levels are present in the nucleus with decreasing levels in other organelles such as cytosol, microsomes and mitochondria (Fig. **2**). It can be written comprehensively as nucleus > cytosol > microsomes > mitochondria [81].

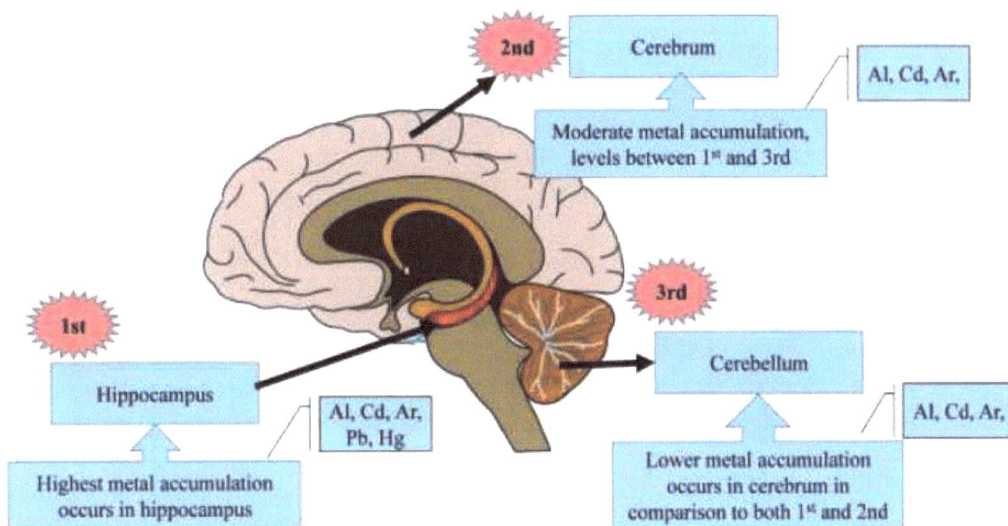

Fig. (2). Comparison of metals accumulation in different brain regions. Labels (1st, 2nd and 3rd) indicate brain regions with the highest, moderate and low levels of toxic metals.

Al is a well-known notorious agent causing learning and memory impairments [82] as Al exposures of brain are linked with decreased neuroplasticity as well as increased risks of neurodegeneration. Memory loss (dementia) is the most common clinical manifestation of Al toxicity in brain. Since hippocampus is the main region of brain which is involved in memory formation as well as neural plasticity [83], so higher levels of Al in the hippocampus are highly defensible with memory loss outcomes [84].

Arsenic

Just like Al, As is also known for its highest concentrations in the hippocampus. But along with hippocampus, it has its equal concentrations in cortical areas followed by the cerebellum (Fig. **2**). Inorganic As is the preferred chemical form in which As gets accumulated in brain parts [85].

Lead

Pb is one of the most toxic metals on earth crust [86]. Pb induced neurotoxicity is due to its compatibility with calcium (Ca) ions. It acts as a substitute of Ca and can get entry through barriers swiftly. Once in the brain, it affects two key processes of brain *i.e.* the cell- to- cell signaling and neurotransmission [87]. The most highly affected areas of brain due to Pb toxicity are the cerebrum, cerebellum and hippocampus [87]. Pb accumulation and damage are not uniform throughout the brain but vary from region to region. The highest percentage of Pb is found in hippocampus followed by both cerebrum and cerebellum (Fig. **2**). If Pb induced shrinkage of brain parts is taken into consideration, then cerebellum is the most affected area [88]. The levels of Pb induced toxicity in different brain regions can be graded from highest to lowest as, hippocampus > cerebrum > cerebellum [89]. The elevated levels of Pb in the brain are a major risk factor in stimulating pathology and progression of various neurological diseases [83].

Mercury

Hg is highly toxic in its methylmercury form [90] and it is the 3rd most toxic metal for brain [53]. The main target of Hg in the CNS is the hippocampus. Deleterious effects of Hg include impaired motor coordination, along with learning and memory impairment [91]. Distribution of and toxic effects of Hg are different in brain's hippocampus, cortex and cerebellum (Fig. **2**). Hippocampus gets most damaged by Hg due to the high accumulation [92].

Cadmium

Cd is also known for its toxicity in adults and in children. It is notorious for causing mental retardation in children, because children brain barriers are not fully developed. Cd is found throughout the brain regions including hippocampus, cortex and cerebellum [93]. Maximum levels of Cd are usually found in choroid plexus as it is the first defense against metals intake from systemic circulation [94].

TOXIC METALS WITH MULTIPLE TOXICOKINETIC ASPECTS

Recognizing the factors which influence toxicity of metals are of wide consideration (Fig. **3**). After getting distributed in different regions of brain, the pharmacokinetic aspects of metals manifests in the form of disrupting calcium signaling, mitochondrial dysfunction, altered neurotransmission, and oxidative stress etc.

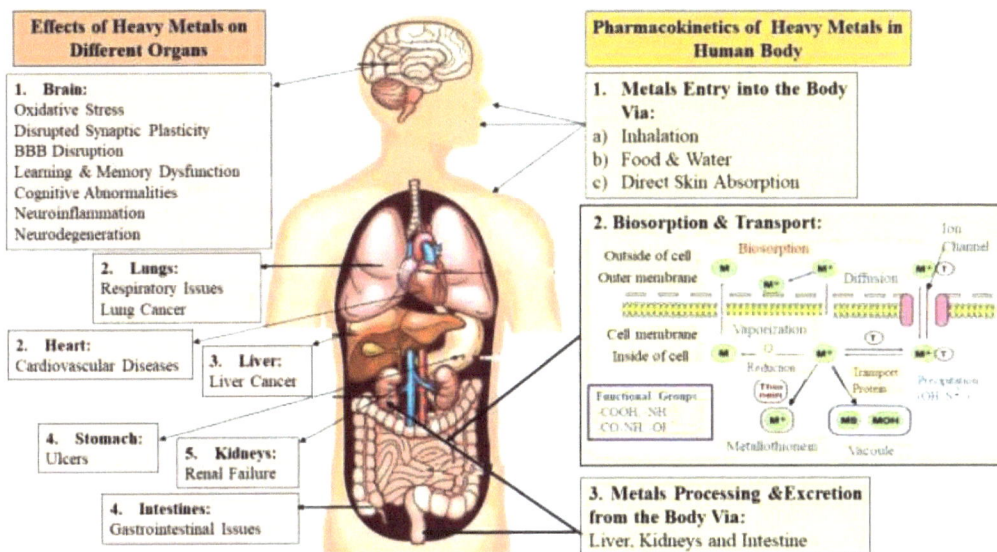

Fig. (3). Pharmacokinetics of Heavy Metals.

Bio-absorption

Among toxic metals, Al is least absorbed dermally. The main route to Al entry is through binding to transferrin protein and once it enters the brain, it can keep the concentration levels up to 1-2 mg/kg of brain [95]. Up to 90% of As gets absorbed by the gastrointestinal tract of human body in the form of arsenite and arsenate. In metabolic pathways, it targets certain proteins and enzymes systems that contain sulfhydryl. Availability of As in brain and other regions depends largely upon the chemical form in which it is absorbed, however, an average half-life of As is about 4 days [96] as metabolites halt the processes in brain by inactivating a series of host enzymes involved in crucial cellular processes [96]. Level of Pb retained in brain vary with age and it is widely accepted that Pb poses its risk more in children than adults. And the reason behind it is that adults only retain 5% of total Pb absorbed and the remaining is excreted [97].

Mitochondrial Dysfunction

Timely generation of the action potential in neurons is necessary for proper signaling. Energy for such robust processes is provided by mitochondria present along the length of axons and throughout the cell body. Oxidative phosphorylation is the key process in regulating the functioning of mitochondria which requires highly regulated membranous systems. Metals toxicity disrupts this process by interfering with the membrane permeability as Al in its ionic form causes excessive leakiness of the membrane [98]. Not only do metals disrupt the membrane permeability but also interact with certain pathways in mitochondria. As interferes with and decreases the activity of mitochondrial complexes I, II-III, and IV. Such alteration can directly lead to disease manifestation such as Parkinson's disease [99]. Mitochondrial swelling is also one of the outcomes of metals toxicity [99]. Ca^{2+} release from the mitochondrial membranes is also compromised due to Pb toxicity. Excess abnormal Ca^{2+} flux causes development of transition pores leading to programmed cell death events [100]. Hg and Cd absorption also leads to the mitochondrial swelling as well as enhanced production of ROS [101].

Effects of Metals Induced Toxicity on Neurotransmission

Human CNS is comprised of around hundred billion neurons. The neurons in the brain are not working individually but are interconnected to one another. This interconnection of neurons is precise and well defined. Communication and coordination of body functions are possible through these neuronal connections. These connection points are called synapses. Synapse is thus the junction between two neurons. Structurally, one neuron can have more than one synaptic connection and, in some cases, even Upton 7000 connections. Presynaptic chemical processes in collaboration with post synaptic signal transduction lead to necessary changes in the brain. Neurotransmitters are the key elements regulating this signal transmission. Any alteration in this process can halt neuron functioning with harmful and destructive disorders ranging from dementia to Parkinson's disease and multiple sclerosis [102]. The presence of toxic metals can affect the process of transmission drastically (Table **3**). In the neurons, toxic metals affect both the signal transmission at the synaptic junction as well as synaptic plasticity resulting in clinical manifestation of impaired learning and memory. Cam kinase pathway is one of the important signaling processes in brain. This pathway is involved in the activation of certain neuronal proteins causing neuronal plasticity. The pathway starts with the Ca^{2+} influx under the influence of certain neurotransmitters mediated by protein, PSD 95. This Ca^+ flux causes activation of another protein, calmodulin which ultimately activates camK protein (signaling protein) [103]. This camK pathway is also affected by certain metals which mimic the Ca^{2+} signaling to block the signaling mechanism [104].

Table 3. Signaling pathways and proteins affected by metals toxicity leading to certain neurological diseases.

Metal	Pathways Blocked	Targets at Protein Level	Diseases	Reference
Aluminium	• Voltage- gated calcium channels • Inhibits Ca^{2+} ATPase	• CaM • CamK • Synapsin • synaptotagmin	• Alzheimer's disease • Neurotoxicity	[82,111]
Cadmium	• Voltage- gated calcium channel • Release of excitatory neurotransmitters	• Adenlycyclate • SNAP25 • ERK1/2 • P38 MAPK	• Mental retardation in children • Apoptosis • Altered gene expression	[112,113]
Lead	• Voltage-gated calcium channel	• GPCR • Adenlycyclate • PKA • CaM	• Schizophrenia • Behavioral problems • Parkinson's disease	[86, 112]
Mercury	• Voltage- gated calcium channels	• Adenlycyclase • Synapsin	• Minamata • Cognitive impairments	[114,115]
Arsenic	• Cholinergic signaling • Glutamatergic signaling • Glucocorticoid signaling	• NMDA receptors • PSD-95 • p-CAMKIIα	• Cognitive impairments • Dementia • Mood disorder	[116,117]

Another strategy adopted by metals is that often more than one metal target these cellular pathways and show a synergistic effect in deteriorating the neuronal structure and function [105]. Talking more specifically, Al shows its toxic effect by blocking the neuronal Ca^{2+} channels [106]. Blockage of Ca^{2+} channels means less signal transmission and less activity of CamK and ultimately, the cellular activity shift from long-term potentiation (LTP) to long-term depression (LTD) (Table **3**). If these conditions prevail, then these may consequently lead to neurodegeneration [107]. Cd toxicity is also manifested in the form of the destruction of Ca^{2+} signaling [108]. Often Cd toxicity is characterized by abnormal transmitter's release. Presence of Cd causes an excessive release of excitatory neurotransmitters, such as, glutamate and aspartate and also an increased release of inhibitory neurotransmitters such as gamma- aminobutyric acid (GABA) [109]. Similarly, As is also found to manifest its toxicity by blocking Ca^{2+} signaling pathway [110].

Cellular Oxidative Stress Induction by Toxic Metals

Toxic metal signaling in brain is carried out importantly by their interactions with the metabolic activities going on in the cellular compartments. Other way to impose their effect is by the production of massive amounts of metals' free radicals in certain parts of the brain, causing neuronal damage by increased oxidative stress [86]. Hence metals are also involved in the production of ROS in the brain (Fig. **4**). The mechanism behind these metals induced oxidative stress is that owing to metals accumulation in the brain there is no more equilibrium between the production and elimination of ROS [118]. At cellular level, oxidative stress is more appropriately defined as a situation when the concentration of ROS is higher. It may be higher at acute or at chronic level with the outcome in the form of impaired cellular metabolic processes and physical damage to cellular components [119]. Adding to the curse of toxic metals, these cause impairment of antioxidant defense. There is an overall unstable and impaired cellular environment which results in damage to important organelles and disrupts biomolecules such as lipids, proteins and sometimes even genetic material [120].

Al and other toxic metals use oxidative stress as the primary pathway for induction of toxicity in the brain. These cause tissue damage in the brain with a high impact on hippocampus and cerebral parts. Damage in parts is carried out by lipid peroxidation, protein depletion, altered gene expression and pathways, and neurodegeneration. An altered expression can have an effect on the signaling pathways in brain such as CamK pathway, which in turn will affect the plasticity, especially LTP [121].

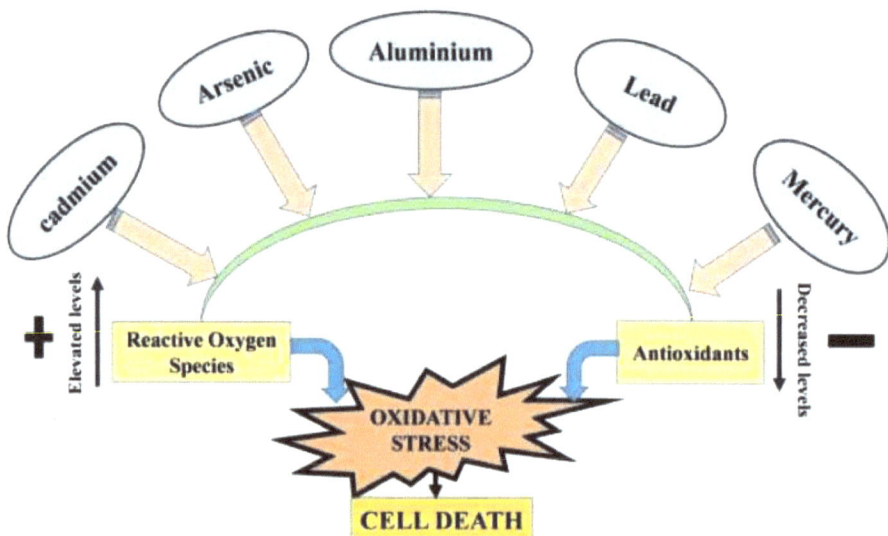

Fig. (4). Metals induced neurotoxicity causes increase in production of ROS and decreased levels of antioxidants, resulting in oxidative stress.

Events of Cellular Signaling in the Nervous System

Most cell signaling occurs through chemicals that may be metallic in nature or neurotransmitters. The most widely distributed chemical in signal transduction is Ca^{2+}. Ca^{2+} signaling in brain maintains the normal processes of synapses. Toxic metals interact with signaling processes by either blocking or decreasing the process or by mimicking the behavior of calcium. Al interferes with the kinetics of cell signaling by altering Ca^{2+} metabolism and increasing its concentration. At secondary level, Al toxicity also interferes with calmodulin which is an essential Ca^{2+} regulator and hence causing huge damage to Ca^{2+} signaling [122]. Similarly, other metals (Hg, As, Cd) also interfere with Ca^{2+} signaling and disrupt the normal process of Ca^{2+} homeostasis [123].

METAL EXCRETION FROM THE BODY

As is mainly excreted through the urine. About 50 - 80% of the consumed inorganic As (having a half-life of approximately 10 hours), is excreted in 3 days. Methylated As takes longer to get excreted from the body as it has a biological half-life of almost 30 hours (three times more than inorganic As). As is also excreted during excessive sweating and by the shedding or removal of skin layer as it is known to have a predilection for the skin [124].

In age 10-15 of the population, the critical concentration for Cd to cause renal dysfunction is approximately 30.2 mg, while in about 50% of the population, the amount is 38.4-42.3 mg. The concentration of Cd increases in the kidney and liver simultaneously, until the final renal cortex concentration reaches 40-50 mg and Cd levels in the liver reach about 70 ppm, depicting a regular pattern [125]. However, this pattern is broken once higher Cd levels are achieved in the liver, thus showing a disproportionate relationship with a metal concentration in the kidney, *i.e.* when then Cd concentration increases in the liver, it significantly decreases in the renal cortex due to excretion [126]. An epidemiological study conducted in the late 1980s showed the dose dependent relationship of Cd intake showed the total consumption of Cd over a period of lifetime that caused health defects and bodily dysfunctions was 2000 mg for both genders [1]. Proteinuria is the presence of protein in the urine at abnormal quantities, and is hence used as an indicator of renal tubular dysfunction. For renal tubular dysfunction, proteinuria mainly consists of proteins with a lower molecular weight due to impaired reabsorption by proximal tubular lining cells, caused by Cd exposure. Till date, the pathogenesis of the glomerular lesion in Cd nephropathy is not understood completely [127].

Once Al enters the body, *via* gastrointestinal absorption, it travels to the liver where, most of it is cleared from the bloodstream [128]. In an experiment conducted on a volunteer (41 yr. old male), it was observed that only 10-15% of Al remained in the body while the rest was excreted the first day. Nonetheless, 7% of Al remained in the body for 170 days after it was initially injected [129]. At the time, the authors had expected its clearance half-life to be less than a year. Pharmacokinetic analysis of the study's results showed the terminal clearance half-life of Al in blood to be 7 days, and whole body to be 170 days. However, in contrast to these results due to sample size taken into consideration, another study [130] determined biliary excretion accounts for less than 1% of Al excretion while excretion *via* urine accounts about 9-17%. Within two weeks, more than 80% of the Al was excreted *via* urine and less than 2% was detected in the feces [129].

At least two kinetic pools can be devised for Pb, which have different turnover rates [131]. The first one is the largest pool which is also the slowest with respect to its kinetics is the bones or skeleton, which has a half-life of more than 20 years. The soft tissue pool is subjected to change most easily, for example Pb found in the trabecular bone is more liable, with a shorter turn over time than that found in the cortical bone [132]. Pb found in the bone may be responsible for about 50% of lead exposure to blood contributing as a major source of internal exposure.

Pb concentration in the bone keeps increasing with age from about 70% in childhood unto about 95% of the body burden with advancing age. Over a lifetime, Pb can accumulate in the body ranging from 200 mg to 500 mg (for a worker who is exposed to the metal on a daily basis). Pb is usually excreted from the body *via* glomerular filtrate in the kidney. however, the metal is also sometimes excreted *via* other route, with other bodily fluids, such as milk during lactation [133].

Hg once absorbed in the tissue, is converted from elemental form to divalent form with the help of catalases. In case Hg is inhaled is vapor form, it gets absorbed in the RBCs and also gets converted to divalent Hg, however, some of the Hg gets transferred to distal areas, *e.g.* brain, as metallic Hg, where its biotransformation may occur [124]. Hg may get excreted from the body *via* urine and feces, depending on the dose of metal received, form of Hg, and time lapsed after exposure. After exposure to inorganic Hg, fecal route is the predominant route of its excretion, however, excretion through urine increases with time [134].

The biological half-lives of Hg are not very well known; however, an estimate has been made of 44 days. The elimination of Hg can be determined by employing more complex kinetics, after exposure to vapors of Hg or its inorganic salts. While the half-lives of Hg are different for various tissues, at times, more than one half life is required to determine the elimination kinetics. Nonetheless, the range of these half-lives is in between 20 to 90 days [124].

DISEASE MANIFESTATION AS AN OUTCOME TO TOXIC METAL EXPOSURE

Alzheimer's Disease (AD)

Alzheimer's is a disease related to the process of aging and is highly progressive, common and fatal. The clinical symptoms of AD include difficulty in performing day to day tasks, and normal activities along with increased memory loss. The manifestation of these symptoms is sometimes presented at the age of 60 years. This is termed as early onset AD (EOAD), which consists of a small portion of disease related cases. As a result of these mutations, amyloid plaques accumulate which is a hallmark for AD. The manifestation of these symptoms may also occur in a later age; termed as late onset AD (LOAD). In LOAD, the cases are usually sporadic, and the manifestation of symptoms occurs at an age later than 60. Increased exposure to Al is linked with morbid changes in the CNS as it leads to the formation of neurofibrillary tangles [135], β-amyloid plaques [136, 137], accumulation of tau protein and neuronal dysfunction [138]. Tau, a microtubule associated protein, needs phosphate for functioning. In case the protein gets hyperphosphorylated, it leads to loss of its biological activity and results in the development of neurofibrillary tangles in patients suffering from AD.

Having a ligand exchange rate, as compared to the other positively charged ions, such as Mg^{2+}, Al also inhibits the enzymes with Mg2+ as a co-factor along with inhibition of processes that involve Ca^{+2} exchange. As a result of these inhibitions, Al causes necrosis of neuronal cells and glial cells. Increased chronic exposure to Al also results in impairment of LTP [139]. Al also causes disruption to the voltage gated Ca^{2+} channels and neurotransmitter receptors, along with impairment of synaptic transmission. Lastly, Al exposure results in deficit of spatial memory, influences emotional reactions of an individual and impairment of many brain functions which are involved with learning and memory [140].

Parkinson's Disease (PD)

PD is the most common disorder of the basal ganglia. It is a progressive neuro-degenerative disease that occurs at an older age. The clinical attributes of PD include resting tremors, bradykinesia, and postural disturbance along with rigidity. However, the manifestation of these attributes only occurs at a loss of about 70-80% of the dopaminergic neurons [141]. These clinical attributes occur due to loss of dopaminergic neurons in the substantia nigra pars compacta, followed by an inherent decrease in dopamine levels in the striatum [142]. Metals induced oxidative stress results in degeneration of dopaminergic neurons [143]. PD is also linked to exposure of brain to heavy metals (such as Cd) and essential elements (such as zinc) with a higher affinity towards sulfhydryl group, which leads to a decrease in D2 dopamine receptor sites. Affinities with the sulfhydryl groups are important. Cd reduces the receptor site by 40-60%, while Hg almost completely depletes the sites by 100% [144].

Exposure to Al leads to an increase in concentration of a metabolite of dopamine 3,4- dihydroxyphenylacetic acid (DOPAC), in areas of Striatum, which is followed by a non-significant decrease in dopamine levels [145]. It has been determined *via* various studies that Al is responsible for increasing the oxidative stress by enhancing the activity of 6- hydroxydopamine, and hence leading to neurodegeneration in the dopaminergic system, making it suggestive of being a potential risk factor in the development of PD [146].

Amyotrophic Lateral Sclerosis (ALS)

ALS is a neurodegenerative disorder which is characterized by the death and loss of motor neurons in cortical regions. Clinical manifestations of disease include weakening of muscles, death of motor neurons, dementia and in some cases respiratory problem [147].

Al toxicity in brain is associated with the manifestation of ALS. Exact mechanism of action of Al and other metals in manifesting ALS is not clearly understood, but many different factors add up to ALS as exposure to Al for a long time or at higher concentrations results in the development of neurofibrillary accumulations same as AD. A direct damage to motor neurons occurs by deteriorating the axons. Damage to axon halts the process of axonal transport and damages motor function [148]. Hg also has a role in the pathogenesis of disease by causing neurodegeneration. In ALS everything surrounds the instability of motor neurons. Upon crossing BBB, Hg tends to accumulate in corticomotor neurons [149]. At a molecular level, Hg is involved in halting some important neurobiological processes such as synapse transmission. It interacts with the Ca^{2+} signaling pathway. Poisoning of Hg causes a hyper excitation of N- methyl- D- aspartic acid (NMDA) receptors in neurons of the cortex. Abnormality of NMDA receptors causes an increased in flux of Ca^{2+} ions. It causes damage to neuronal cytoskeleton proteins. It also causes induction of oxidative stress and neuroinflammation by interacting with kB pathway [150].

Cd is involved in hindering the normal antioxidant activity by interacting with one of the important protein Copper (Cu), Zinc(Zn)-Superoxide dismutase (SOD1) involved in removing the excess free radicals from the cells. These proteins contain Cu and Zn ions in their structure. Cd can replace these metals form the SOD1 by mimicking their structure and thus aborts the activity of Cu, Zn-SOD protein. In this way, Cd increases the levels of ROS in the cell leading to cases of apoptotic cell death and neurodegeneration [151].

Pb and As are notorious for their ability to accumulate in brain as well as in bones. Etiology of Pb and As causing ALS is not well defined at yet, but it can be easily correlated that Pb is directly involved in increased oxidative stress, and hence it can act as a culprit in worsening the process of neurodegeneration in ALS [152].

Multiple Sclerosis (MS)

MS is a progressive degenerative disorder of CNS with somewhat unknown etiology. It is an autoimmune disease in which the immune cells start attacking the brain cells. The target area for the immune cells is the myelin covering of the neurons. At axons the myelin sheath covers the axon and ensures steady signal transmission. Immune cells attack the myelin sheath and cause its deterioration. Loss of myelin sheath results in poor signal transmission. There is a loss of communication between the brain and the rest of the body. Progression of disease takes place with time and there is no treatment available for MS other than symptomatic relief. Inflammatory events are considered to be the hallmark of this disease pathology along with oxidative stress [153].

Toxic metals such as Al, As, Cd, Hg and Pb are considered to be the toxins in triggering the process of auto immune. These toxic metals along with some other metals, are clear markers for disease [154]. Other than triggering harmful cellular processes, these metals can also deposit on the myelin sheath and causes changes in biomolecules such as proteins and glycoproteins from the membrane structure. In response to this there is the generation of auto antibodies from immune cells as well as initiation of the phagocytic process to destroy these metals but in turn acts as accumulators of plaques on the myelin sheath [155].

Al shows its association with MS. A high level of Al gets accumulated in the human body especially in the CNS [156]. The known association of Al with both myelin and oligodendrocytes might be the reason for a presumed relationship of the metal and the disease. While, the role of Al in the development of the disease along with its progression is currently unknown, it may still be linked to Al activity as a pro-oxidant or its ability as an adjuvant to induce a kind of autoimmunity in affected tissues. Al toxicity caused by either of these possible mechanisms result in breakdown of myelin in MS [157].

As a fatal neurotoxin, Hg has effects on the basal cellular functions, by forming strong bonds with selenohydryl and sulphydryl groups that are found on the albumin molecules within cell membranes, receptors, and intracellular signal bindings. Hg is considered to be toxic as it produces free radicals and causes a possible change in cell redox reactions. Hg is harmful for oligodendroglia of the brain and causes damage to oligodendrocytes *via* apoptosis [158].

MS may also be induced in the neuronal cells by As *via* inflammation, degeneration and apoptosis. Another reason for the disease pathogenesis could be the generation of ROS, in the cells upon exposure to the heavy metal [159]. Cd and Pb also play role in pathogenesis of MS by inducing brain toxicity. Both these metals cause damage to the BBB, thus activate the process of neurodegeneration ultimately leading to pathogenesis of MS [160].

Synergistic Effects of Metals Mixture in Brain

Each year, high loads of toxic metals are released into our environments from sources such as smelting, mining and agriculture. Many heavy metals are a part of our ecosystem and they exist either individually or in the form of mixtures, for example Hg, Pb, Cd and As [161]. A recent concern is the occurrence of these metals in low dose mixtures [162]. Metal exposure has detrimental effects on humans and they may be exposed through accumulation in food, water or environment [163]. Metals like As, Cd, Hg, Pb are toxic in nature and are reported to cause a number of disorders like, neurological conditions, hypertension, cerebral palsy, cognitive impairments, blindness, cirrhosis, hyperkeratosis, dysarthria and cancers [164, 165]. According to studies, individual toxic metal exposure interferes with essential metals in various tissues and can cause toxicity in lungs, kidneys, brain etc [166].

Co- exposure of metals like As, Pb and Mn can result in an increased level of Pb in the brain, due to the presence of high affinity Pb binding proteins in the brain [167]. Current research suggests that toxic metals like Cd, As, Hg and Pb, even at low concentrations may prove detrimental to organisms [168]. However, the research on the interaction toxic metal mixtures with essential metals is still inadequate and requires more attention.

CONCLUDING REMARKS

Metals play important role in human life as many of them are required in trace amounts in certain enzymatic reactions taking place in the body. But excessive of these metals and many other environmental toxic metals including Al, Pb, As, Hg, Cd etc. can lead to detrimental health consequences especially neurological effects. Many neurological disorders like AD, PD, ALS, MS, and Attention deficit hyperactivity disorder (ADHD) etc. have been linked to direct overexposure of toxic metals. Pharmacokinetic parameters of these metals are crucial to be discussed. The obvious favored paths for these metals to get entry into brain is through BBB and BCB by causing structural damage or by disrupting the functioning of choroid plexuses. Toxic metals adopt certain strategies to hinder the normal pharmacokinetic regulation in the brain such as disrupting calcium signaling, mitochondrial dysfunction, biotransformation and especially inducing

ROS and impairing neuronal transmission, resulting in structural and functional damage to neurons across the CNS. This process increases with increasing age as the chances of metals exposure also increase, but more studies are still required in order to understand the combined effects of these metals on the brain, their specific individual pathways of action and the potential therapeutic targets to be used to block these pathways.

CONSENT FOR PUBLICATION

Not Applicable.

CONFLICT OF INTEREST

The author declares no conflict of interest, financial or otherwise.

ACKNOWLEDGEMENTS

Declared none.

ABBREVIATIONS

AD	Alzheimer's Disease
ADHD	Attention Deficit Hyper Activity Disorder
ALS	Amyotropic Lateral Sclerosis
Al	Aluminium
As	Arsenic
BBB	Blood Brain Barrier
BCB	Blood Cerebrospinal Fluid Barrier
BDNF	Brain Derived Neurotropic Factor
Ca^{2+}	Calcium
CamK	Calcium/Calmodulin Dependent Protein Kinase
Cd	Cadmium
CECs	Cerebral Endothelial Cells
CP	Choroid Plexus
CSF	Cerebrospinal Fluid
CNS	Central Nervous System
Cu	Copper
CVDs	Cardiovascular Diseases
DOPAC	3,4- Dihydroxyphenylacetic Acid
EOAD	Early Onset of Alzheimer's Disease
ERK	Extracellular Signal- Regulated Kinase

GABA	Gamma Amino Butyric Acid
GIT	Gastrointestinal Tract
GPCR	G- Protein Coupled Receptor
Hg	Mercury
LOAD	Late Onset of Alzheimer's Disease
LTD	Long- term Depression
LTP	Long- term Potentiation
MAPK	Mitogen- Activated Protein Kinase
MS	Multiple Sclerosis
NMDA	N- methyl- d- aspartate
Pb	Lead
PD	Parkinson's Disease
PKA	Protein Kinase A
PSD95	Post Synaptic Density Protein 95
ROS	Reactive Oxygen Species
SNAP	Synaptosomal- Associated Protein
SOD	Super Oxide Dismutase
TJs	Tight Junctions
Zn	Zinc

REFERENCES

[1]   Järup L, Berglund M, Elinder CG, Nordberg G, Vanter M. Health effects of cadmium exposure–a review of the literature and a risk estimate. Scand J Work Environ Health 1998; 24: 1-52.

[2]   Schlesinger WH. Biogeochemistry. Elsevier 2005.

[3]   Maret W. The Metals in the Biological Periodic System of the Elements: Concepts and Conjectures. Int J Mol Sci 2016; 17(1): 66.
[http://dx.doi.org/10.3390/ijms17010066] [PMID: 26742035]

[4]   Monosson E. Evolution in a toxic world: How life responds to chemical threats. Island Press 2012.
[http://dx.doi.org/10.5822/978-1-61091-221-1]

[5]   Anastassopoulou J, Theophanides T. The Role of Metal Ions in Biological Systems and Medicine.Bioinorganic Chemistry: An Inorganic Perspective of Life. Dordrecht: Springer Netherlands 1995; pp. 209-18.
[http://dx.doi.org/10.1007/978-94-011-0255-1_17]

[6]   Slater JP, Tamir I, Loeb LA, Mildvan AS. The mechanism of Escherichia coli deoxyribonucleic acid polymerase I. Magnetic resonance and kinetic studies of the role of metals. J Biol Chem 1972; 247(21): 6784-94.
[http://dx.doi.org/10.1016/S0021-9258(19)44655-3] [PMID: 4343158]

[7]   Brini M, Ottolini D, Calì T, Carafoli E. Calcium in health and disease Interrelations between essential metal ions and human diseases. Springer 2013; pp. 81-137.
[http://dx.doi.org/10.1007/978-94-007-7500-8_4]

[8]   Moore SE, Logerwell E, Eisner L, Farley EV, Harwood LA, Kuletz K, *et al.* Marine fishes, birds and mammals as sentinels of ecosystem variability and reorganization in the Pacific Arctic region The Pacific arctic region. Springer 2014; pp. 337-92.

[9]   Khadilkar A, Khadilkar V, Chinnappa J, *et al.* Prevention and treatment of vitamin D and calcium deficiency in children and adolescents: Indian Academy of Pediatrics (IAP). Guidelines Indian Pediatrics 2017; 54(7): 567-73.
[http://dx.doi.org/10.1007/s13312-017-1070-x] [PMID: 28737142]

[10]   Shugart HA. Heavy viewing: emergent frames in contemporary news coverage of obesity. Health Commun 2011; 26(7): 635-48.
[http://dx.doi.org/10.1080/10410236.2011.561833] [PMID: 21541866]

[11]   Rakova N. Long-term studies on sodium balance and their relevance to human health 2016.

[12]   Medicine Dietary Reference Intakes for sodium and potassium. National Academies Press 2019.

[13]   Mancia G, Oparil S, Whelton PK, *et al.* The technical report on sodium intake and cardiovascular disease in low- and middle-income countries by the joint working group of the World Heart Federation, the European Society of Hypertension and the European Public Health Association. Eur Heart J 2017; 38(10): 712-9.
[http://dx.doi.org/10.1093/eurheartj/ehw549] [PMID: 28110297]

[14]   van der Wijst J, Tutakhel OAZ, Bos C, *et al.* Effects of a high-sodium/low-potassium diet on renal calcium, magnesium, and phosphate handling. Am J Physiol Renal Physiol 2018; 315(1): F110-22.
[http://dx.doi.org/10.1152/ajprenal.00379.2017] [PMID: 29357414]

[15]   Zoroddu MA, Aaseth J, Crisponi G, Medici S, Peana M, Nurchi VM. The essential metals for humans: a brief overview. J Inorg Biochem 2019; 195: 120-9.
[http://dx.doi.org/10.1016/ j.jinorgbio.2019.03.013] [PMID: 30939379]

[16]   Bertinato J, Wu Xiao C, Ratnayake WM, *et al.* Lower serum magnesium concentration is associated with diabetes, insulin resistance, and obesity in South Asian and white Canadian women but not men. Food Nutr Res 2015; 59(1): 25974.
[http://dx.doi.org/10.3402 /fnr.v59.25974] [PMID: 25947295]

[17]   Gröber U, Schmidt J, Kisters K. Magnesium in prevention and therapy. Nutrients 2015; 7(9): 8199-226.
[http://dx.doi.org/10.3390 /nu 7095388] [PMID: 26404370]

[18]   Jahnen-Dechent W, Ketteler M. Magnesium basics. Clin Kidney J 2012; 5 (Suppl. 1): i3-i14.
[http://dx.doi.org/10.1093 /ndtplus/sfr163] [PMID: 26069819]

[19]   Martin KJ, González EA, Slatopolsky E. Clinical consequences and management of hypomagnesemia. J Am Soc Nephrol 2009; 20(11): 2291-5.
[http://dx.doi.org/10.1681/ASN.2007111194] [PMID: 18235082]

[20]   Abbaspour N, Hurrell R, Kelishadi R. Review on iron and its importance for human health. J Res Med Sci 2014; 19(2): 164-74.
[PMID: 24778671]

[21]   Al-Fartusie FS, Mohssan SN. Essential trace elements and their vital roles in human body. Indian J Adv Chem Sci 2017; 5(3): 127-36.

[22]   Chellan P, Sadler PJ. The elements of life and medicines. Philos Trans- Royal Soc, Math Phys Eng Sci 2015; 373(2037)20140182
[http://dx.doi.org/10.1098/rsta.2014.0182] [PMID: 25666066]

[23]   Chen P, Bornhorst J, Aschner M. Manganese metabolism in humans 2019.

[24]   Gong JH, Lo K, Liu Q, *et al.* Dietary Manganese, Plasma Markers of Inflammation, and the Development of Type 2 Diabetes in Postmenopausal Women: Findings From the Women's Health Initiative. Diabetes Care 2020; 43(6): 1344-51.

[http://dx.doi.org/10.2337/dc20-0243] [PMID: 32295807]

[25] Rahal A, Shivay YS. Micronutrient deficiencies in humans and animals: strategies for their improvement Biofortification of Food Crops. Springer 2016; pp. 217-28.

[26] Hara T, Takeda TA, Takagishi T, Fukue K, Kambe T, Fukada T. Physiological roles of zinc transporters: molecular and genetic importance in zinc homeostasis. J Physiol Sci 2017; 67(2): 283-301.
[http://dx.doi.org/10.1007/s12576-017-0521-4] [PMID: 28130681]

[27] Gu HF, Zhang X. Zinc Deficiency and Epigenetics 2017.
[http://dx.doi.org/10.1007/978-3-319-40007-5_80-1]

[28] Ogawa Y, Kawamura T, Shimada S. Zinc and skin biology. Arch Biochem Biophys 2016; 611: 113-9.
[http://dx.doi.org/10.1016/j.abb.2016.06.003] [PMID: 27288087]

[29] Kaim W, Schwederski B, Klein A. Bioinorganic Chemistry--Inorganic Elements in the Chemistry of Life: An Introduction and Guide. John Wiley & Sons 2013.

[30] Solomon EI. Spectroscopic methods in bioinorganic chemistry: blue to green to red copper sites. Inorg Chem 2006; 45(20): 8012-25.
[http://dx.doi.org/10.1021/ic060450d] [PMID: 16999398]

[31] Messerschmidt A. Handbook of metalloproteins. Wiley 2001.

[32] Steffens GC, Soulimane T, Wolff G, Buse G. Stoichiometry and redox behaviour of metals in cytochrome-c oxidase. Eur J Biochem 1993; 213(3): 1149-57.
[http://dx.doi.org/10.1111/j.1432-1033.1993.tb17865.x] [PMID: 8389295]

[33] Shils ME, Shike M. Modern nutrition in health and disease. Lippincott Williams & Wilkins 2006.

[34] Bustos RI, Jensen EL, Ruiz LM, *et al.* Copper deficiency alters cell bioenergetics and induces mitochondrial fusion through up-regulation of MFN2 and OPA1 in erythropoietic cells. Biochem Biophys Res Commun 2013; 437(3): 426-32.
[http://dx.doi.org/10.1016/j.bbrc.2013.06.095] [PMID: 23831624]

[35] Barton RA. Neocortex size and behavioural ecology in primates. Proc Biol Sci 1996; 263(1367): 173-7.
[http://dx.doi.org/10.1098/rspb.1996.0028] [PMID: 8728982]

[36] Crichton R, Ward R. Metal-based neurodegeneration: from molecular mechanisms to therapeutic strategies. John Wiley & Sons 2013.
[http://dx.doi.org/10.1002/9781118553480]

[37] Takeda A. [Essential trace metals and brain function]. Yakugaku Zasshi 2004; 124(9): 577-85.
[http://dx.doi.org/10.1248 /yakushi.124.577] [PMID: 15340179]

[38] Dabrowiak JC. Metals in medicine. John Wiley & Sons 2017.
[http://dx.doi.org/10.1002/9781119191377]

[39] Elscnhans B, Schümann K, Forth W. Toxic metals: interactions with essential metals. Boston, London: Nutrition, Toxicity and Cancer CRC Press, Boca Raton, An Arbor 1991; pp. 223-58.

[40] Engwa GA, Ferdinand PU, Nwalo FN, Unachukwu MN. Mechanism and health effects of heavy metal toxicity in humans Poisoning in the Modern World-New Tricks for an Old Dog?. IntechOpen 2019.

[41] Arain MS, Afridi HI, Kazi TG, *et al.* Variation in the Levels of Aluminium and Manganese in Scalp Hair Samples of the Patients Having Different Psychiatric Disorders with Related to Healthy Subjects. Biol Trace Elem Res 2015; 168(1): 67-73.
[http://dx.doi.org/10.1007/s12011-015-0353-0] [PMID: 25947935]

[42] Kramer M. Sleep loss in resident physicians: the cause of medical errors? Front Neurol 2010; 1: 128.
[http://dx.doi.org/10.3389/fneur.2010.00128] [PMID: 21188260]

[43] Barnhisel R, Bertsch PM. Aluminium. Methods of Soil Analysis: Part 2 Chemical and Microbiological

Properties 1983; 9: 275-300.

[44]   Shakir SK, Azizullah A, Murad W, Daud MK, Nabeela F, Rahman H, *et al.* Toxic metal pollution in Pakistan and its possible risks to public health Reviews of environmental contamination and toxicology. Springer 2016; pp. 1-60.

[45]   US Environmental Protection Agency2020

[46]   World Health Organization2020

[47]   Gary F. Common heavy metals: sources and specific effects 2012.

[48]   Abernathy CO, Thomas DJ, Calderon RL. Health effects and risk assessment of arsenic. J Nutr 2003; 133(5) (Suppl. 1): 1536S-8S.
[http://dx.doi.org/10.1093/jn/133.5.1536S] [PMID: 12730460]

[49]   Brammer H, Ravenscroft P. Arsenic in groundwater: a threat to sustainable agriculture in South and South-east Asia. Environ Int 2009; 35(3): 647-54.
[http://dx.doi.org/10.1016/j.envint.2008.10.004] [PMID: 19110310]

[50]   Gidlow DA. Lead toxicity. Occup Med (Lond) 2004; 54(2): 76-81.
[http://dx.doi.org/10.1093/occmed/kqh019] [PMID: 15020724]

[51]   Riess ML, Halm JK. Lead poisoning in an adult: lead mobilization by pregnancy? J Gen Intern Med 2007; 22(8): 1212-5.
[http://dx.doi.org/10.1007/s11606-007-0253-x] [PMID: 17562116]

[52]   Ullah R, Malik RN, Qadir A. Assessment of groundwater contamination in an industrial city, Sialkot, Pakistan. Afr J Environ Sci Technol 2009; 3(12)

[53]   Rice KM, Walker EM Jr, Wu M, Gillette C, Blough ER. Environmental mercury and its toxic effects. J Prev Med Public Health 2014; 47(2): 74-83.
[http://dx.doi.org/10.3961/jpmph.2014.47.2.74] [PMID: 24744824]

[54]   Marettová E, Maretta M, Legáth J. Toxic effects of cadmium on testis of birds and mammals: a review. Anim Reprod Sci 2015; 155: 1-10.
[http://dx.doi.org/10.1016/j.anireprosci.2015.01.007] [PMID: 25726439]

[55]   Neal A, Guilarte T. Mechanisms of heavy metal neurotoxicity: lead and manganese. J Drug Metab Toxicol S 2012; 5(2)

[56]   Campbell A, Becaria A, Lahiri DK, Sharman K, Bondy SC. Chronic exposure to aluminium in drinking water increases inflammatory parameters selectively in the brain. J Neurosci Res 2004; 75(4): 565-72.
[http://dx.doi.org/10.1002/jnr.10877] [PMID: 14743440]

[57]   Gulya K, Rakonczay Z, Kása P. Cholinotoxic effects of Aluminium in rat brain. J Neurochem 1990; 54(3): 1020-6.
[http://dx.doi.org/10.1111/j.1471-4159.1990.tb02352.x] [PMID: 2303807]

[58]   Ghribi O, Herman MM, Forbes MS, DeWitt DA, Savory J. GDNF protects against Aluminium-induced apoptosis in rabbits by upregulating Bcl-2 and Bcl-XL and inhibiting mitochondrial Bax translocation. Neurobiol Dis 2001; 8(5): 764-73.
[http://dx.doi.org/10.1006/nbdi.2001.0429] [PMID: 11592846]

[59]   Kawahara M, Kato-Negishi M, Hosoda R, Imamura L, Tsuda M, Kuroda Y. Brain-derived neurotrophic factor protects cultured rat hippocampal neurons from Aluminium maltolate neurotoxicity. J Inorg Biochem 2003; 97(1): 124-31.
[http://dx.doi.org/10.1016/S0162-0134(03)00255-1] [PMID: 14507468]

[60]   Rodríguez VM, Jiménez-Capdeville ME, Giordano M. The effects of arsenic exposure on the nervous system. Toxicol Lett 2003; 145(1): 1-18.
[http://dx.doi.org/10.1016/S0378-4274(03)00262-5] [PMID: 12962969]

[61]   Tyler CR, Allan AM. The effects of arsenic exposure on neurological and cognitive dysfunction in human and rodent studies: a review. Curr Environ Health Rep 2014; 1(2): 132-47.
[http://dx.doi.org/10.1007/s40572-014-0012-1] [PMID: 24860722]

[62]   Flora SJ, Flora G, Saxena G. Environmental occurrence, health effects and management of lead poisoning. Lead: Elsevier 2006; pp. 158-228.
[http://dx.doi.org/10.1016/B978-044452945-9/50004-X]

[63]   Flora G, Gupta D, Tiwari A. Toxicity of lead: A review with recent updates. Interdiscip Toxicol 2012; 5(2): 47-58.
[http://dx.doi.org/10.2478/v10102-012-0009-2] [PMID: 23118587]

[64]   A Review of: "Medical Toxicology" AU - Brent, Jeffrey. Clin Toxicol 2006; 44(3): 355.

[65]   Bellinger D. Lead Pediatrics Find this article online 2004; 113: 1016-22.

[66]   Fernandes Azevedo B, Barros Furieri L, Peçanha FM, Wiggers GA, Frizera Vassallo P, Ronacher Simões M, *et al.* Toxic effects of mercury on the cardiovascular and central nervous systems. BioMed Research International 2012.
[http://dx.doi.org/10.1155/2012/949048]

[67]   McNutt M. Mercury and health. American Association for the Advancement of Science 2013.
[http://dx.doi.org/10.1126/science.1245924]

[68]   Burukoğlu D, Bayçu C. Protective effects of zinc on testes of cadmium-treated rats. Bull Environ Contam Toxicol 2008; 81(6): 521-4.
[http://dx.doi.org/10.1007/s00128-007-9211-x] [PMID: 18925378]

[69]   Smith QR, Rabin O, Chikhale EG. Delivery of metals to brain and the role of the blood-brain barrier Metals and oxidative damage in neurological disorders. Springer 1997; pp. 113-30.
[http://dx.doi.org/10.1007/978-1-4899-0197-2_7]

[70]   Zheng W, Aschner M, Ghersi-Egea J-F. Brain barrier systems: a new frontier in metal neurotoxicological research. Toxicol Appl Pharmacol 2003; 192(1): 1-11.
[http://dx.doi.org/10.1016/S0041-008X(03)00251-5] [PMID: 14554098]

[71]   Ballabh P, Braun A, Nedergaard M. The blood-brain barrier: an overview: structure, regulation, and clinical implications. Neurobiol Dis 2004; 16(1): 1-13.
[http://dx.doi.org/10.1016/j.nbd.2003.12.016] [PMID: 15207256]

[72]   Lai C-H, Kuo K-H, Leo JM. Critical role of actin in modulating BBB permeability. Brain Res Brain Res Rev 2005; 50(1): 7-13.
[http://dx.doi.org/10.1016/j.brainresrev.2005.03.007] [PMID: 16291072]

[73]   Wolburg H, Lippoldt A. Tight junctions of the blood-brain barrier: development, composition and regulation. Vascul Pharmacol 2002; 38(6): 323-37.
[http://dx.doi.org/10.1016/S1537-1891(02)00200-8] [PMID: 12529927]

[74]   Yokel RA. Blood-brain barrier flux of Aluminium, manganese, iron and other metals suspected to contribute to metal-induced neurodegeneration. J Alzheimers Dis 2006; 10(2-3): 223-53.
[http://dx.doi.org/10.3233/JAD-2006-102-309] [PMID: 17119290]

[75]   Bhowmik A, Khan R, Ghosh MK. Blood brain barrier: a challenge for effectual therapy of brain tumors. BioMed research international 2015.
[http://dx.doi.org/10.1155/2015/320941]

[76]   Saunders NR, Dreifuss J-J, Dziegielewska KM, *et al.* The rights and wrongs of blood-brain barrier permeability studies: a walk through 100 years of history. Front Neurosci 2014; 8: 404.
[http://dx.doi.org/10.3389/fnins.2014.00404] [PMID: 25565938]

[77]   Wright RO, Baccarelli A. Metals and neurotoxicology. J Nutr 2007; 137(12): 2809-13.
[http://dx.doi.org/10.1093/jn/137.12.2809] [PMID: 18029504]

[78]   Vinceti M, Filippini T, Mandrioli J, *et al.* Lead, cadmium and mercury in cerebrospinal fluid and risk of amyotrophic lateral sclerosis: A case-control study. J Trace Elem Med Biol 2017; 43: 121-5.
[http://dx.doi.org/10.1016/j.jtemb.2016.12.012] [PMID: 28089071]

[79]   Karri V, Schuhmacher M, Kumar V. Heavy metals (Pb, Cd, As and MeHg) as risk factors for cognitive dysfunction: A general review of metal mixture mechanism in brain. Environ Toxicol Pharmacol 2016; 48: 203-13.
[http://dx.doi.org/10.1016/j.etap.2016.09.016] [PMID: 27816841]

[80]   Zheng W. Toxicology of choroid plexus: special reference to metal-induced neurotoxicities. Microsc Res Tech 2001; 52(1): 89-103.
[http://dx.doi.org/10.1002/1097-0029(20010101)52:1<89::AID-JEMT11>3.0.CO;2-2]   [PMID: 11135452]

[81]   Julka D, Vasishta RK, Gill KD. Distribution of Aluminium in different brain regions and body organs of rat. Biol Trace Elem Res 1996; 52(2): 181-92.
[http://dx.doi.org/10.1007/BF02789460] [PMID: 8773759]

[82]   Mehpara Farhat S, Mahboob A, Ahmed T. Oral exposure to Aluminium leads to reduced nicotinic acetylcholine receptor gene expression, severe neurodegeneration and impaired hippocampus dependent learning in mice. Drug Chem Toxicol 2019; •••: 1-9.
[http://dx.doi.org/10.1080/01480545.2019.1587452] [PMID: 30889993]

[83]   Iqbal G, Zada W, Mannan A, Ahmed T. Elevated heavy metals levels in cognitively impaired patients from Pakistan. Environ Toxicol Pharmacol 2018; 60: 100-9.
[http://dx.doi.org/10.1016/j.etap.2018.04.011] [PMID: 29684799]

[84]   Ganrot PO. Metabolism and possible health effects of Aluminium. Environ Health Perspect 1986; 65: 363-441.
[PMID: 2940082]

[85]   Li J, Duan X, Dong D, *et al.* Tissue-specific distributions of inorganic arsenic and its methylated metabolites, especially in cerebral cortex, cerebellum and hippocampus of mice after a single oral administration of arsenite. J Trace Elem Med Biol 2017; 43: 15-22.
[http://dx.doi.org/10.1016/j.jtemb.2016.10.002] [PMID: 27745987]

[86]   Bhatti S, Ali Shah SA, Ahmed T, Zahid S. Neuroprotective effects of Foeniculum vulgare seeds extract on lead-induced neurotoxicity in mice brain. Drug Chem Toxicol 2018; 41(4): 399-407.
[http://dx.doi.org/10.1080/01480545.2018.1459669] [PMID: 29742941]

[87]   Sanders T, Liu Y, Buchner V, Tchounwou PB. Neurotoxic effects and biomarkers of lead exposure: a review. Rev Environ Health 2009; 24(1): 15-45.
[http://dx.doi.org/10.1515/REVEH.2009.24.1.15] [PMID: 19476290]

[88]   Sidhu P, Nehru B. Lead intoxication: histological and oxidative damage in rat cerebrum and cerebellum. Journal of Trace Elements in Experimental Medicine: The Official Publication of the International Society for Trace Element Research in Humans 2004; 17(1): 45-53.
[http://dx.doi.org/10.1002/jtra.10052]

[89]   Fjerdingstad EJ, Danscher G, Fjerdingstad E. Hippocampus: selective concentration of lead in the normal rat brain. Brain Res 1974; 80(2): 350-4.
[http://dx.doi.org/10.1016/0006-8993(74)90699-4] [PMID: 4422673]

[90]   Clarkson TW, Magos L. The toxicology of mercury and its chemical compounds. Crit Rev Toxicol 2006; 36(8): 609-62.
[http://dx.doi.org/10.1080/10408440600845619] [PMID: 16973445]

[91]   Drew MR, Huckleberry KA. Modulation of aversive memory by adult hippocampal neurogenesis. Neurotherapeutics 2017; 14(3): 646-61.
[http://dx.doi.org/10.1007/s13311-017-0528-9] [PMID: 28488160]

[92]   Sakamoto M, Nakano A. Comparison of mercury accumulation among the brain, liver, kidney, and the

brain regions of rats administered methylmercury in various phases of postnatal development. Bull Environ Contam Toxicol 1995; 55(4): 588-96.
[http://dx.doi.org/10.1007/BF00196040] [PMID: 8555685]

[93]   Rigon AP, Cordova FM, Oliveira CS, *et al.* Neurotoxicity of cadmium on immature hippocampus and a neuroprotective role for p38 MAPK. Neurotoxicology 2008; 29(4): 727-34.
[http://dx.doi.org/10.1016/j.neuro.2008.04.017] [PMID: 18541302]

[94]   Wang B, Du Y. Cadmium and its neurotoxic effects. Oxidative medicine and cellular longevity 2013.
[http://dx.doi.org/10.1155/2013/898034]

[95]   Weisser K, Stübler S, Matheis W, Huisinga W. Towards toxicokinetic modelling of aluminium exposure from adjuvants in medicinal products. Regul Toxicol Pharmacol 2017; 88: 310-21.
[http://dx.doi.org/10.1016/j.yrtph.2017.02.018] [PMID: 28237896]

[96]   Mochizuki H. Arsenic neurotoxicity in humans. Int J Mol Sci 2019; 20(14): 3418.
[http://dx.doi.org/10.3390/ijms20143418] [PMID: 31336801]

[97]   Abadin H, Taylor J, Buser MC, Scinicariello F, Przybyla J, Klotzbach JM, *et al.* Toxicological profile for lead: draft for public comment 2019.

[98]   Iglesias-González J, Sánchez-Iglesias S, Beiras-Iglesias A, Méndez-Álvarez E, Soto-Otero R. Effects of aluminium on rat brain mitochondria bioenergetics: an *in vitro* and in vivo study. Mol Neurobiol 2017; 54(1): 563-70.
[http://dx.doi.org/10.1007/s12035-015-9650-z] [PMID: 26742531]

[99]   Zorov DB, Juhaszova M, Sollott SJ. Mitochondrial reactive oxygen species (ROS) and ROS-induced ROS release. Physiol Rev 2014; 94(3): 909-50.
[http://dx.doi.org/10.1152/physrev.00026.2013] [PMID: 24987008]

[100] Brookes PS, Yoon Y, Robotham JL, Anders MW, Sheu S-S. Calcium, ATP, and ROS: a mitochondrial love-hate triangle. Am J Physiol Cell Physiol 2004; 287(4): C817-33.
[http://dx.doi.org/10.1152/ajpcell.00139.2004] [PMID: 15355853]

[101] Kahrizi F, Salimi A, Noorbakhsh F, *et al.* Repeated administration of mercury intensifies brain damage in multiple sclerosis through mitochondrial dysfunction. Iranian journal of pharmaceutical research. Iran J Pharm Res 2016; 15(4): 834-41.
[PMID: 28243280]

[102] Kutter EF, Bostroem J, Elger CE, Mormann F, Nieder A. Single neurons in the human brain encode numbers. Neuron 2018.
[http://dx.doi.org/10.1016/j.neuron.2018.08.036]

[103] Rand J, Nonet M. Synaptic transmission 1997.

[104] Shioda N, Fukunaga K. Physiological and pathological roles of CaMKII-PP1 signaling in the brain. Int J Mol Sci 2017; 19(1): 20.
[http://dx.doi.org/10.3390/ijms19010020] [PMID: 29271887]

[105] Nday CM, Drever BD, Salifoglou T, Platt B. Aluminium interferes with hippocampal calcium signaling in a species-specific manner. J Inorg Biochem 2010; 104(9): 919-27.
[http://dx.doi.org/10.1016/j.jinorgbio.2010.04.010] [PMID: 20510457]

[106] Amber S, Shah SAA, Ahmed T, Zahid S. Syzygium aromaticum ethanol extract reduces AlCl3-induced neurotoxicity in mice brain through regulation of amyloid precursor protein and oxidative stress gene expression. Asian Pac J Trop Med 2018; 11(2): 123.
[http://dx.doi.org/10.4103/1995-7645.225019]

[107] Cataldi M. The changing landscape of voltage-gated calcium channels in neurovascular disorders and in neurodegenerative diseases. Curr Neuropharmacol 2013; 11(3): 276-97.
[http://dx.doi.org/10.2174/1570159X11311030004] [PMID: 24179464]

[108] Yuan Y, Jiang CY, Xu H, *et al.* Cadmium-induced apoptosis in primary rat cerebral cortical neurons

culture is mediated by a calcium signaling pathway. PLoS One 2013; 8(5)e64330
[http://dx.doi.org/10.1371/journal.pone.0064330] [PMID: 23741317]

[109] Minami A, Takeda A, Nishibaba D, Takefuta S, Oku N. Cadmium toxicity in synaptic neurotransmission in the brain. Brain Res 2001; 894(2): 336-9.
[http://dx.doi.org/10.1016/S0006-8993(01)02022-4] [PMID: 11251212]

[110] Splettstoesser F, Florea AM, Büsselberg D. IP(3) receptor antagonist, 2-APB, attenuates cisplatin induced Ca2+-influx in HeLa-S3 cells and prevents activation of calpain and induction of apoptosis. Br J Pharmacol 2007; 151(8): 1176-86.
[http://dx.doi.org/10.1038/sj.bjp.0707335] [PMID: 17592515]

[111] Hashmi AN, Yaqinuddin A, Ahmed T. Pharmacological effects of Ibuprofen on learning and memory, muscarinic receptors gene expression and APP isoforms level in pre-frontal cortex of AlCl₃-induced toxicity mouse model. Int J Neurosci 2015; 125(4): 277-87.
[http://dx.doi.org/10.3109/00207454.2014.922972] [PMID: 24825584]

[112] Iqbal G, Ahmed T. Co-exposure of metals and high fat diet causes aging like neuropathological changes in non-aged mice brain. Brain Res Bull 2019; 147: 148-58.
[http://dx.doi.org/10.1016/j.brainresbull.2019.02.013] [PMID: 30807793]

[113] Branca JJV, Morucci G, Pacini A. Cadmium-induced neurotoxicity: still much ado. Neural Regen Res 2018; 13(11): 1879-82.
[http://dx.doi.org/10.4103/1673-5374.239434] [PMID: 30233056]

[114] lorianne Monnet-Tschudi F, Boschat C, Corbaz A, Honegger P. Involvement of environmental mercury and lead in the etiology of neurodegenerative diseases. Rev Environ Health 2006; 21(2)

[115] Korogi Y, Takahashi M, Okajima T, Eto K. MR findings of Minamata disease--organic mercury poisoning. J Magn Reson Imaging 1998; 8(2): 308-16.
[http://dx.doi.org/10.1002/jmri.1880080210] [PMID: 9562057]

[116] Srivastava P, Dhuriya YK, Gupta R, *et al.* Protective effect of curcumin by modulating BDNF/DARPP32/CREB in arsenic-induced alterations in dopaminergic signaling in rat corpus striatum. Mol Neurobiol 2018; 55(1): 445-61.
[http://dx.doi.org/10.1007/s12035-016-0288-2] [PMID: 27966075]

[117] Xi S, Sun W, Wang F, Jin Y, Sun G. Transplacental and early life exposure to inorganic arsenic affected development and behavior in offspring rats. Arch Toxicol 2009; 83(6): 549-56.
[http://dx.doi.org/10.1007/s00204-009-0403-5] [PMID: 19212760]

[118] Losacco C, Perillo A. Metal-Induced Oxidative Stress and Cellular Signaling Alteration in Animals. Iran J Appl Anim Sci 2018; 8(3): 367-73.

[119] Pandey G, Madhuri S. Heavy metals causing toxicity in animals and fishes. Research Journal of Animal. Veterinary and Fishery Sciences 2014; 2(2): 17-23.

[120] Winterbourn CC. Reconciling the chemistry and biology of reactive oxygen species. Nat Chem Biol 2008; 4(5): 278-86.
[http://dx.doi.org/10.1038/nchembio.85] [PMID: 18421291]

[121] Sharma B, Singh S, Siddiqi NJ. Biomedical implications of heavy metals induced imbalances in redox systems. BioMed research international 2014.
[http://dx.doi.org/10.1155/2014/640754]

[122] Zhang J, Liu S, Zhang L, Nian H, Chen L. Effect of Aluminium stress on the expression of calmodulin and the role of calmodulin in Aluminium tolerance. J Biosci Bioeng 2016; 122(5): 558-62.
[http://dx.doi.org/10.1016/j.jbiosc.2016.04.001] [PMID: 27133707]

[123] Roos D, Seeger R, Puntel R, Vargas Barbosa N. Role of calcium and mitochondria in MeHg-mediated cytotoxicity. Journal of Biomedicine and Biotechnology 2012.
[http://dx.doi.org/10.1155/2012/248764]

[124] Goyer RA, Clarkson TW. Toxic effects of metals. Casarett and Doull's toxicology: the basic science of poisons 1996; 5: 696-8.

[125] Ellis KJ, Yuen K, Yasumura S, Cohn SH. Dose-response analysis of cadmium in man: body burden *vs* kidney dysfunction. Environ Res 1984; 33(1): 216-26.
[http://dx.doi.org/10.1016/0013-9351(84)90018-5] [PMID: 6363055]

[126] Nogawa K, Honda R, Kido T, *et al.* A dose-response analysis of cadmium in the general environment with special reference to total cadmium intake limit. Environ Res 1989; 48(1): 7-16.
[http://dx.doi.org/10.1016/S0013-9351(89)80080-5] [PMID: 2644119]

[127] Liu F, Jan K-Y. DNA damage in arsenite- and cadmium-treated bovine aortic endothelial cells. Free Radic Biol Med 2000; 28(1): 55-63.
[http://dx.doi.org/10.1016/S0891-5849(99)00196-3] [PMID: 10656291]

[128] Xu ZX, Tang JP, Badr M, Melethil S. Kinetics of Aluminium in rats. III: Effect of route of administration. J Pharm Sci 1992; 81(2): 160-3.
[http://dx.doi.org/10.1002/jps.2600810212] [PMID: 1545356]

[129] Priest N, Newton D, Talbot R. Metabolism of Aluminium-26 and gallium-67 in a volunteer following their injection as citrates. UKAEA Harwell report AEA-EE-0206 1991.

[130] Xu ZX, Pai SM, Melethil S. Kinetics of Aluminium in rats. II: Dose-dependent urinary and biliary excretion. J Pharm Sci 1991; 80(10): 946-51.
[http://dx.doi.org/10.1002/jps.2600801009] [PMID: 1784003]

[131] O'Flaherty EJ. Physiologically based models of metal kinetics. Crit Rev Toxicol 1998; 28(3): 271-317.
[http://dx.doi.org/10.1080/10408449891344209] [PMID: 9631283]

[132] Silbergeld EK, Schwartz J, Mahaffey K. Lead and osteoporosis: mobilization of lead from bone in postmenopausal women. Environ Res 1988; 47(1): 79-94.
[http://dx.doi.org/10.1016/S0013-9351(88)80023-9] [PMID: 3168967]

[133] Gulson BL, Mahaffey KR, Jameson CW, *et al.* Mobilization of lead from the skeleton during the postnatal period is larger than during pregnancy. J Lab Clin Med 1998; 131(4): 324-9.
[http://dx.doi.org/10.1016/S0022-2143(98)90182-2] [PMID: 9579385]

[134] Miettinen J. Absorption and elimination of dietary mercury (Hg2+) and methylmercury in man. Mercury: Mercurials and Mercaptans 1973.

[135] Vasudevaraju P, Govindaraju M, Palanisamy AP, Sambamurti K, Rao KS. Molecular toxicity of aluminium in relation to neurodegeneration. Indian J Med Res 2008; 128(4): 545-56.
[PMID: 19106446]

[136] Rodella LF, Ricci F, Borsani E, *et al.* Aluminium exposure induces Alzheimer's disease-like histopathological alterations in mouse brain. Histol Histopathol 2008; 23(4): 433-9.
[PMID: 18228200]

[137] Praticò D. Oxidative imbalance and lipid peroxidation in Alzheimer's disease. Drug Dev Res 2002; 56(3): 446-51.
[http://dx.doi.org/10.1002/ddr.10097]

[138] Terada S, Sato S, Nagao S, *et al.* Trail making test B and brain perfusion imaging in mild cognitive impairment and mild Alzheimer's disease. Psychiatry Res Neuroimaging 2013; 213(3): 249-55.
[http://dx.doi.org/10.1016/j.pscychresns.2013.03.006] [PMID: 23830931]

[139] Díaz-Nido J, Avila J. Aluminium induces the *in vitro* aggregation of bovine brain cytoskeletal proteins. Neurosci Lett 1990; 110(1-2): 221-6.
[http://dx.doi.org/10.1016/0304-3940(90)90815-Q] [PMID: 2109292]

[140] Kawahara M, Kato-Negishi M. Link between Aluminium and the pathogenesis of Alzheimer's disease: the integration of the Aluminium and amyloid cascade hypotheses. International journal of Alzheimer's disease 2011.

[http://dx.doi.org/10.4061/2011/276393]

[141] Forno LS. Pathological considerations in the etiology of Parkinson's disease. NEUROLOGICAL DISEASE AND THERAPY 1995; 40: 65.

[142] Dexter DT, Wells FR, Lees AJ, *et al.* Increased nigral iron content and alterations in other metal ions occurring in brain in Parkinson's disease. J Neurochem 1989; 52(6): 1830-6.
[http://dx.doi.org/10.1111/j.1471-4159.1989.tb07264.x] [PMID: 2723638]

[143] Jenner P. Oxidative mechanisms in nigral cell death in Parkinson's disease. Mov Disord 1998; 13 (Suppl. 1): 24-34.
[PMID: 9613715]

[144] Jankovic J. Searching for a relationship between manganese and welding and Parkinson's disease. Neurology 2005; 64(12): 2021-8.
[http://dx.doi.org/10.1212/01.WNL.0000166916.40902.63] [PMID: 15985567]

[145] Li J-Y, Englund E, Holton JL, *et al.* Lewy bodies in grafted neurons in subjects with Parkinson's disease suggest host-to-graft disease propagation. Nat Med 2008; 14(5): 501-3.
[http://dx.doi.org/10.1038/nm1746] [PMID: 18391963]

[146] Sánchez-Iglesias S, Méndez-Alvarez E, Iglesias-González J, *et al.* Brain oxidative stress and selective behaviour of aluminium in specific areas of rat brain: potential effects in a 6-OHDA-induced model of Parkinson's disease. J Neurochem 2009; 109(3): 879-88.
[http://dx.doi.org/10.1111/j.1471-4159.2009.06019.x] [PMID: 19425176]

[147] Kiernan MC, Vucic S, Cheah BC, *et al.* Amyotrophic lateral sclerosis. Lancet 2011; 377(9769): 942-55.
[http://dx.doi.org/10.1016/S0140-6736(10)61156-7] [PMID: 21296405]

[148] Tanridag T, Coskun T, Hürdag C, Arbak S, Aktan S, Yegen B. Motor neuron degeneration due to aluminium deposition in the spinal cord: a light microscopical study. Acta Histochem 1999; 101(2): 193-201.
[http://dx.doi.org/10.1016/S0065-1281(99)80018-X] [PMID: 10335362]

[149] Pamphlett R, Kum Jew S. Uptake of inorganic mercury by human locus ceruleus and corticomotor neurons: implications for amyotrophic lateral sclerosis. Acta Neuropathol Commun 2013; 1(1): 13.
[http://dx.doi.org/10.1186/2051-5960-1-13] [PMID: 24252585]

[150] Cariccio VL, Samà A, Bramanti P, Mazzon E. Mercury involvement in neuronal damage and in neurodegenerative diseases. Biol Trace Elem Res 2019; 187(2): 341-56.
[http://dx.doi.org/10.1007/s12011-018-1380-4] [PMID: 29777524]

[151] Huang YH, Shih CM, Huang CJ, *et al.* Effects of cadmium on structure and enzymatic activity of Cu,Zn-SOD and oxidative status in neural cells. J Cell Biochem 2006; 98(3): 577-89.
[http://dx.doi.org/10.1002/jcb.20772] [PMID: 16440303]

[152] Ingre C, Roos PM, Piehl F, Kamel F, Fang F. Risk factors for amyotrophic lateral sclerosis. Clin Epidemiol 2015; 7: 181-93.
[PMID: 25709501]

[153] Ontaneda D, Thompson AJ, Fox RJ, Cohen JA. Progressive multiple sclerosis: prospects for disease therapy, repair, and restoration of function. Lancet 2017; 389(10076): 1357-66.
[http://dx.doi.org/10.1016/S0140-6736(16)31320-4] [PMID: 27889191]

[154] Stankiewicz J, Panter SS, Neema M, Arora A, Batt CE, Bakshi R. Iron in chronic brain disorders: imaging and neurotherapeutic implications. Neurotherapeutics 2007; 4(3): 371-86.
[http://dx.doi.org/10.1016/j.nurt.2007.05.006] [PMID: 17599703]

[155] Fulgenzi A, Zanella SG, Mariani MM, Vietti D, Ferrero ME. A case of multiple sclerosis improvement following removal of heavy metal intoxication: lessons learnt from Matteo's case. Biometals 2012; 25(3): 569-76.
[http://dx.doi.org/10.1007/s10534-012-9537-7] [PMID: 22438029]

[156] Exley C, Mamutse G, Korchazhkina O, *et al.* Elevated urinary excretion of aluminium and iron in multiple sclerosis. Mult Scler 2006; 12(5): 533-40.
[http://dx.doi.org/10.1177/1352458506071323] [PMID: 17086897]

[157] Jones K, Linhart C, Hawkins C, Exley C. Urinary excretion of aluminium and silicon in secondary progressive multiple sclerosis. EBioMedicine 2017; 26: 60-7.
[http://dx.doi.org/10.1016/j.ebiom.2017.10.028] [PMID: 29128442]

[158] Attar AM, Kharkhaneh A, Etemadifar M, Keyhanian K, Davoudi V, Saadatnia M. Serum mercury level and multiple sclerosis. Biol Trace Elem Res 2012; 146(2): 150-3.
[http://dx.doi.org/10.1007/s12011-011-9239-y] [PMID: 22068727]

[159] Alizadeh-Ghodsi M, Zavvari A, Ebrahimi-Kalan A, Shiri-Shahsavar MR, Yousefi B. The hypothetical roles of arsenic in multiple sclerosis by induction of inflammation and aggregation of tau protein: A commentary. Nutr Neurosci 2018; 21(2): 92-6.
[http://dx.doi.org/10.1080/1028415X.2016.1239399] [PMID: 27697018]

[160] Aliomrani M, Sahraian MA, Shirkhanloo H, Sharifzadeh M, Khoshayand MR, Ghahremani MH. Blood concentrations of cadmium and lead in multiple sclerosis patients from Iran. Iranian journal of pharmaceutical research. Iran J Pharm Res 2016; 15(4): 825-33.
[PMID: 28243279]

[161] Liu J, Chen L, Cui H, Zhang J, Zhang L, Su C-Y. Applications of metal-organic frameworks in heterogeneous supramolecular catalysis. Chem Soc Rev 2014; 43(16): 6011-61.
[http://dx.doi.org/10.1039/C4CS00094C] [PMID: 24871268]

[162] Luo J, Hendryx M. Relationship between blood cadmium, lead, and serum thyroid measures in US adults - the National Health and Nutrition Examination Survey (NHANES) 2007-2010. Int J Environ Health Res 2014; 24(2): 125-36.
[http://dx.doi.org/10.1080/09603123.2013.800962] [PMID: 23782348]

[163] Maleci L, Buffa G, Wahsha M, Bini C. Morphological changes induced by heavy metals in dandelion (Taraxacum officinale Web.) growing on mine soils. J Soils Sediments 2014; 14(4): 731-43.
[http://dx.doi.org/10.1007/s11368-013-0823-y]

[164] Boucher DS, Platts W, Eds. Prospecting for native metals in lunar polar craters. 7th Symposium on Space Resource Utilization.
[http://dx.doi.org/10.2514/6.2014-0338]

[165] He J, Chen JP. A comprehensive review on biosorption of heavy metals by algal biomass: materials, performances, chemistry, and modeling simulation tools. Bioresour Technol 2014; 160: 67-78.
[http://dx.doi.org/10.1016/j.biortech.2014.01.068] [PMID: 24630371]

[166] Cobbina SJ, Chen Y, Zhou Z, *et al.* Low concentration toxic metal mixture interactions: Effects on essential and non-essential metals in brain, liver, and kidneys of mice on sub-chronic exposure. Chemosphere 2015; 132: 79-86.
[http://dx.doi.org/10.1016/j.chemosphere.2015.03.013] [PMID: 25828250]

[167] Andrade V, Mateus ML, Batoréu MC, Aschner M, dos Santos APM. Urinary delta-ALA: a potential biomarker of exposure and neurotoxic effect in rats co-treated with a mixture of lead, arsenic and manganese. Neurotoxicology 2013; 38: 33-41.
[http://dx.doi.org/10.1016/j.neuro.2013.06.003] [PMID: 23764341]

[168] Huang M-Y, Duan R-Y, Ji X. The influence of long-term cadmium exposure on phonotaxis in male Pelophylax nigromaculata. Chemosphere 2015; 119: 763-8.
[http://dx.doi.org/10.1016/j.chemosphere.2014.08.014] [PMID: 25192651]

CHAPTER 2

Co-Exposure of Aluminium with other Metals Causes Neurotoxicity and Neurodegeneration

Abida Zulfiqar[1], Ghazal Iqbal[1] and Touqeer Ahmed[1,*]

[1] *Neurobiology Laboratory, Department of Healthcare Biotechnology, Atta-ur-Rahman School of Applied Biosciences, National University of Sciences and Technology, Sector H-12, Islamabad 44000, Pakistan*

Abstract: Metals are key players in maintaining and regulating gene expression, antioxidant response, cell structure and neurotransmission. Their presence in the human body is required in trace amounts to perform these functions, however, excessive accumulation of these metals in various organs, including the brain, leads to detrimental neurological consequences by altering oxidative stress, protein misfolding, mitochondrial dysfunction, DNA fragmentation and apoptosis. These events over a course of time contribute to mild cognitive impairment, movement related disorders, learning and memory deficits which can further progress to neurodegeneration. According to some epidemiological and clinical findings, there is strong evidence of metal exposure and its correlation with a number of neurological diseases like Alzheimer's diseases (AD), Huntington's disease (HD), amyotrophic lateral sclerosis (ALS), Guillain-Barre disease (GBD), Parkinson's disease (PD) and multiple sclerosis (MS), etc. Moreover, metal ions tend to exacerbate the accumulation of neurotic plaques in AD associated pathologies. It has been observed that metals like iron, zinc, copper and Aluminium are elevated in AD brains, causing damage to the synapses. Such metal ions imbalances are associated with aging related neuropathies and disease progression. Some other factors contributing to neurodegeneration include predisposition to ApoE allele, the interaction and synergistic effect of multiple metals together, the impact of cholesterol, amyloid precursor protein (APP) processing, and increased total tau along with Aβ production play a key role in increased biosynthesis of reactive oxygen species in the brain. Such events tend to reduce neuronal viability and function, thus causing cognitive decline.

Keywords: Metals, Mild cognitive Impairment, Neurodegeneration, Neurotoxicity.

** Corresponding author Touqeer Ahmed:** Neurobiology Laboratory, Department of Healthcare Biotechnology, Atta-ur-Rahman School of Applied Biosciences, National University of Sciences and Technology, Sector H-12, Islamabad - 44000, Pakistan; Tel:+ 92-51-9085-6141, Fax:+ 92-51-9085-6102; E-mail: touqeer.ahmed@asab.nust.edu.pk

1. ACCUMULATION OF METALS IN THE BRAIN AND COGNITIVE IMPAIRMENT

Cognitive impairment is a collective term used when a person's cognitive functions are compromised. There is an overall deficit in learning and memory because such people have difficulty in remembering things, their language, visuospatial abilities are also impaired [1]. This altogether accounts for an increased disability risk along with substantial costs of health care [2]. Cognitive functions are also affected by other factors like environment, genetics, food consumption/diet, life style and aging [3]. With aging, a decline in cognitive functions is inevitable. This is mostly related to an overall increase in oxidative stress and abnormal aggregation of proteins [4]. Moreover, there are other risk factors that can accelerate the process of aging, these include smoking, alcohol consumption, depression, a western or high fat diet, pollution etc [5]. The cognitive deficit is increasing on a global level and it is expected to increase significantly more in developing countries. Along with these above mentioned factors, exposure to other neurotoxins cause increased oxidative stress and affects cognitive functions which then leads to irreversible neurodegeneration [6].

One of the most important risk factors for cognitive decline is exposure to heavy metals from the environment. The human body cannot synthesize or destroy metals. Therefore, in order to provide essential trace elements such as Copper (Cu) and Zinc (Zn) to the body, there are efficient mechanisms for transport, absorption and cellular uptake. Mostly heavy metals utilize this mechanism of transport to enter and accumulate in the various organs in the body [7]. These accumulated metals start the production of reactive oxygen species (nitric oxide and superoxide) in the biological system and result in a surge in oxidative stress. Increased reactive oxygen species (ROS) induce various DNA damages, protein modifications and lipid peroxidation resulting in neurotoxicity and neurodegeneration [8]. Oxidative stress worsens the cognitive abilities by damaging the functions of endothelial cells and invasion of macrophages to the brain parenchyma resulting in disruption of the blood brain barrier. This results in deficit of nutrients in brain cells and as a consequence neuroinflammatory processes begin [9]. Table 1 shows that how elevated levels of some metals can cause certain disorders.

Table 1. Elevated levels of heavy metals and different diseases associated along with their permissible limits.

Metal	Environmental Sources	Diseases Associated with Higher Metal Levels	Nervous System and Related Disorders	Permissible Limits of Metals in Body*	Safe Intake Quantities of Metal in Drinking Water*
Cu	Meat, fish [10]	Bone disorder [11], hepatitis [12], cancer, cardiovascular disease [13]	Motor neuron diseases [14], Alzheimer's disease [15]	0.1 mg/L 0.5µg/kg	1ppm
Al	Beverages can and foil, cosmetics, antiperspirants, antacids and water treatment plants [16]	Ischemic heart disease [17], chronic kidney disease [18], pulmonary dysfunction [19]	Alzheimer's disease [20], Parkinson's disease [21], dementia [22]	0.1µg/kg	0.1ppm
Zn	Food [23], zinc fumes, soil [24], Refineries.	Cancer, cardiovascular disease [13], diabetes mellitus.	Alzheimer's Disease [25]	15mg/L 1mg/kg	5ppm
Pb	Contaminated food, lead containing paint, drinking-water, lead-glazed material for food storage and petrol.	Developmental disabilities [26]	Neurotoxicity, attention deficit hyperactivity disorder, risk of developing mild mental retardation and lead encephalopathy [26, 27]	0.1mg/L 25µg/kg	0.05ppm
Mn	Diet, drinking water [28], metal laden dust in industrial areas [29]	Anemia [30], diabetes mellitus [31]	Parkinson's disease [32], neurotoxicity [33]	0.26mg/L 0.06mg/kg	0.4ppm
Cd	Color pigment present in paints, fertilizers, fabrication of nickel-cadmium batteries, anticorrosive agent, stabilizer in PVC products, tobacco smoking, and used as neutron-absorber in nuclear power plants [34].	Kidney damage, disturbed calcium and phosphorus metabolism, a possible higher risk of kidney stones [35], bone damage [36] and cancer [37].	Motor neuron disease, *e.g.* Amyotrophic lateral sclerosis [38]	0.06mg/L 7µg/kg	0.005ppm

ppm part per million, µg/kg = microgram per kilogram, mg/L = milligram per liter;(World Health Organization, 2008).

2. METALS AND HFD INDUCED OXIDATIVE STRESS IN COGNITIVE IMPAIRMENT

Globalization has led to an increase in lifestyle changes which is directly linked with the consumption of unhealthy food. Processed and fatty food has become readily available and food preferences have changed over the course of time from simpler to more complex [39]. People are more inclined towards consumption of a high fat diet containing carbohydrates and meat in less developed countries due to low level of awareness and education and therefore are more prone to heart diseases, hypertension, diabetes and neurological disorders [40]. Prolonged intake of a high fat diet is a key risk factor for the onset of neurodegenerative functions and causes hippocampus induced memory loss [41]. Moreover, HFD consumption increases oxidative stress and results in a cascade of inflammatory processes in the brain [42]. It also results in lowering the levels of brain derived neurotrophic factors, which play a vital role in the growth, overall support and survival of neurons along with the expression of different antioxidant proteins [43].

Metals also interfere with glutamatergic and cholinergic neurotransmission and, as a result cause cognitive decline. HFD tends to affect the cholinergic neurotransmission by lowering the acetylcholine levels in brain, which leads to a decline in memory loss and cognitive abilities [44, 45]. Many studies have used aged rats and mice as models for studying neuropathological changes and it was concluded based on these studies that there is a potential role of metals in concert with HFD in causing aging like symptoms in young mice [9, 46, 47].

3. PREVALENCE AND CURRENT SCENARIO OF COGNITIVE IMPAIRMENT IN WORLDWIDE

Cognitive impairment is emerging as one of the serious health issues, and the number of individuals is constantly increasing worldwide [48]. It is a common disease among aging individuals and there is a dramatic rise in a number of people living in old homes due to better facilities and thus results in increased life expectancy and decreased mortality rates. This situation leads to an overall increase in cognitive impairment prevalence rates worldwide [49]. In America alone, around 5.1 million (2.5%) old aged individuals suffer from different cognitive disorders and about 10 million American families are providing for cognitively impaired people [50]. According to the epidemiological data, there is a prevalence of about 3% to 19% of mild cognitive impairment worldwide in the elderly population and about 50% of such cases progress to moderate or severe cognitive loss within 3 to 5 years [51, 52].

In Pakistan, people are constantly exposed to a number of heavy metals from dietary and environmental sources. A study has found that elevated levels of heavy metals like Zn, Hg, Cu, Cd, Ni, Pb, and Mn in drinking water throughout different regions of Pakistan and these metal levels surpass the standards set by WHO significantly [53]. The main sources of metal contamination are textile, fertilizer and pharmaceutical industries, along with the crops irrigated with metal water [9, 54]. Therefore, in many developing countries these metal pollutants add to cognitive impairment and other neurological diseases like Alzheimer's, dementia and amyloid plagues by increasing oxidative stress [9, 55].

4. ALUMINIUM NEUROTOXICITY AND ITS EFFECT ON NEUROTRANSMISSION

Aluminium (Al) is one of the highly abundant elements which exists in only (+3) oxidation state, and does not readily go through reduction reactions. It can, however, form complexes when it reacts with some other metals in the environment [56]. People are exposed to this toxic metal mainly through water, food and air and once it gains entry into the human body, about 10% of the Al is absorbed as well as accumulated in the kidney, bones, liver and brain [57]. It has been documented that chronic exposure to Al for longer periods, leads to cognitive impairment and neurodegeneration along with other acute and chronic diseases [58]. According to studies, it is speculated that Al forms a complex called Al-superoxide anion, which is more strong as compared to the superoxide anion alone and forms hydroxyl radicals along with hydrogen peroxide, thus contributing to increased oxidative stress [59].

Of all the organs, brain is particularly sensitive to oxidative stress induced by prolonged Al exposure which results in decreased antioxidant potential and increased level of free radicals. This condition creates a neurotoxic environment and causes neurophysical, neuropathological and neurobehavioral changes [60]. Constant Al exposure leads to its accumulation in various brain regions like corpus callosum, cingulate bundle, medial striatum and most importantly in the hippocampus which cause memory and learning deficits [61, 62]. Neurodegenerative disorders may arise as a consequence of mitochondrial dysfunction. The most important role of mitochondria is energy dependent metabolism, any defect in generation of ATP leads to cellular dysfunction, energy failure and eventually cell death as observed in numerous neurodegenerative disorders [63]. Moreover, the degeneration of mitochondria is also an important and primitive sign of Alzheimer's disease even before the neurofibrillary tangles are visible [64].

The role of Aluminium in neurotoxicity is evident from an Al-based AD model, it has been established that Al exposure leads to hyper-phosphorylated tau, oxidative damage and granulo-vacuolar degeneration in rats. Tau is a well-known microtubule associated protein that can function in the presence of phosphate. If tau gets hyper-phosphorylated, it loses its biological activity and forms neurofibrillary tangles [65]. It was also found that Al concentration was elevated in patients of Parkinson's disease (PD), specifically in the dopamine related brain regions, showing a connection between Al exposure and PD [66, 67]. Al is capable of causing oxidative stress by enhancing the 6-hydroxydopamine and cause neurodegeneration in the dopaminergic systems, which makes it a strong risk factor in PD pathogenesis [66]. According to another study, chronic exposure of Al increased the Aβ42 production by facilitating APP expression as well as beta and gamma secretases in rat models [68]. Al is also reported to induce lipid peroxidation (LPO) in rat brain, which occurs as a consequence of ROS and reactive nitrogen species (RNS) production [69]. The toxic effects of Al include nerve cell atrophy, proliferation of microglia and astrocytes in the thalamus and cerebral cortex. At a dose as low as 100 nM, and its exposure for 3.5 days, Al sulfate tend to stimulate pro-apoptotic and pro-inflammatory gene expression in a primary culture of human brain cells [70].

Al alters the neurotransmission in two ways, either by blocking the enzymes responsible for synthesis of neurotransmitter directly or by changing the mechanism of uptake/release of these neurotransmitters by disturbing physical properties of synaptic membranes [71, 72].

5. MOLECULAR BASIS OF ALUMINIUM NEUROTOXICITY

Aluminium and Cell Signaling

Although there is inadequate information about the Al cell signaling in the brain, it is reported that many bio-molecules such as hormones, neuromodulators and neurotransmitters exert their action through an intermediate system of second messengers. This cell signaling pathway is reported to be blocked by Al both *in vitro* and in vivo systems by interacting with receptor associated G-protein (RAG), membrane receptors, phosphatidylinositol-specific phospholipase C (PI-PLC) or their other substrates [71]. Al tend to bind to polyphosphoinositides through its negative charge and results in lipid clustering. An increase in polyphosphoinositides local concentration as well as incomplete neutralization of negative charge after binding of Al, results in an impaired binding of PL-PLC to its substrates and hence decrease the activity of enzyme. Moreover, oligodendrocytes do not accumulate Al in the absence of transferrin, and show a high metabolism of PIP2 upon low concentration of Al exposure. All the above mentioned findings suggest that Al affects the signaling cascade by altering the binding of regulatory proteins [73].

According to a study, Al forms a lipophilic complex with maltolate, and it has been found to occur in the gastrointestinal tract in vivo. This Al-maltolate complex is itself neurotoxic and causes multiple cytoskeletal abnormalities that can exacerbate the formation of neurofibrillary tangles and eventually apoptosis takes place. Al-maltolate cause apoptosis in the N2a neuroblastoma cells by upregulation of a transcription factor p53 and Bax (apoptosis regulator protein) as well as downregulated expression of Bcl-2 which is an anti-apoptotic protein through intrinsic pathway [74]. Neurodegeneration is usually caused by tissue inflammation, Al exposure increases the expression of NF-κB and TNF-α which creates a neuroinflammatory environment in human glioblastoma cells. The NF-κb and TNF-α activation in turn promotes the downstream expression of iNOS, cytokines and complement factors. These events promote cell death and tissue damage by activation and proliferation of glial cells as shown in Fig. (1) [75].

Neurotoxicity induced by Metals leading to neuronal dysfunction

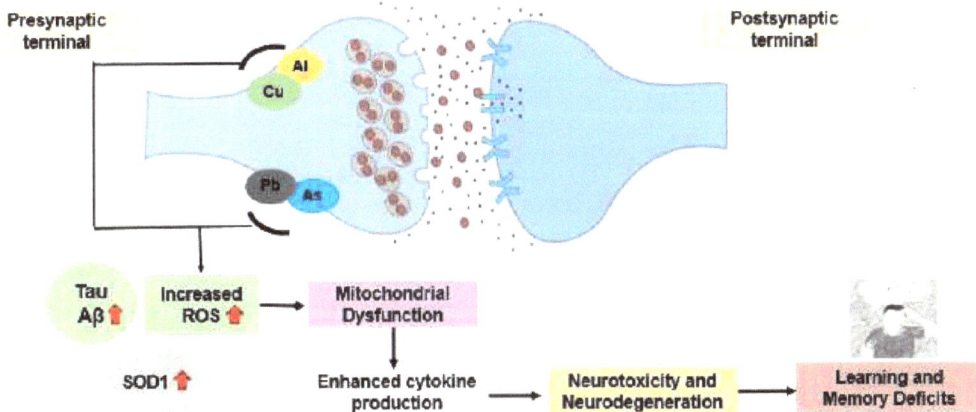

Fig. (1). Neurotoxicity induced by Metals leading to neuronal dysfunction. Diagrammatic representation of metal inducted neurotoxicity leading to neurodegenerative diseases. Metals like Aluminium (Al), Arsenic (As), Lead (Pb), Copper (Cu) etc. cause oxidative stress due to increased levels of (Superoxide dismutase1) SOD1 which also results in tau hyperphosphorylation by the increased expression of MAPK, CDK5 etc. This contributes towards overproduction and aggregation of Aβ, mitochondrial damages leading to synaptic dysfunction and ultimately causing learning and memory deficits. Moreover, the increase in tau protein leads to an increase in cytokine production, thus activating many inflammatory pathways like the iNOS by Ca^{2+} influx through the NMDAR, thereby exacerbating the process of tau hyperphosphorylation.

The mitogen activated protein kinases (MAPKs) also play a critical role in Al neurotoxicity. In the neuronal cell culture system, Al triggered the c-jun N-terminal kinase pathway (SAPK/JNK), an important stimulator of AP-1, a transcription factor required in regulation of apoptosis and cell proliferation. Moreover, in the human neuronal cells, Al also upregulates the levels of hypoxia induced factor (HIF-1) [70]. These findings altogether suggest that Al has either a direct or indirect outcome on expression of MAPK/AP-1, NF-κB, and HIF-1 genes, which altogether increase oxidative stress and lead to neuronal decay. Furthermore, decrease of the anti-inflammatory molecules like the brain derived growth factor and neurotrophins nerve growth factor and stimulation of proinflammatory signals contribute to the process of neuroinflammation in association with Al deposition in different areas of the brain [76].

6. THE NEUROTOXICITY INDUCED BY METAL MIXTURE

As we are continuously being exposed to an environment loaded with different heavy metals, it is noteworthy to study the effects of different metals alone and in combination with others. It has been suggested that different metals act in concert with one another, *i.e.* the level of one metal may have a significant effect on the activity of another metal, and such changes affect the homeostasis of the blood brain barrier (BBB) [77]. Although there is limited research regarding the combined effect of metals, it is however suggested that various metals like Mn, Zn, Cu, Cd, Fe and Zn are carried through shared metal carriers/transporters and have coinciding signaling pathways [78].

Of all the metals, the neurotoxicity of lead (Pb) with other heavy metals like Mn, Hg, Cd and As are well studied. According to a research, prenatal exposure to As and Pb together, raises the chances of intellectual frailty as compared to single metal exposure. Children having high levels of prenatal Cd exposure face severe mental and psychomotor developmental issues if they are exposed to another heavy metal like Pb at some point in life. Moreover, co-exposure of metals like Mn and Pb together in the prenatal period leads to language development issues and cognitive deficits. Pb was also found to be associated with lower IQ score in children who already had higher Mn blood levels. It is important to know and study more that how different metals work together and produce a synergetic effect [79].

It has also been reported that either co-administration or exposure to two/multiple different metals intensifies redistribution and retention of a single metal. Co-exposure of Hg and Se elevated the metal retention of these metals and also caused Hg redistribution in organs and blood [80]. According to a study by Chandra *et al*., intraperitoneal administration of Pb and oral intake of Mn in rats lead to impaired learning and motor abilities. Moreover, the weight of rat brain decreased to a greater extent under co-exposure of Pb and Mn both than either metal alone [81].

The As and Pb interaction has also been reported to bring alterations in the central monaminergic systems in mice. It was found that Pb accumulated significantly but As accumulation was reduced in the brain upon As/Pb exposure. The serotonin levels were found to be increased in the midbrain and frontal cortex, and decreased norepinephrine levels in the hippocampal region when compared with single metal exposure. The metal mix (Mn, As, Pb) significantly reduced rat motor activity when compared to single metal exposure. It was found that the mining waste composed of metal mix (As, Mn, Pb, Cd), when consumed by rats, showed accumulation of Mn and As in brain [79]. Al, may not necessarily produce an oxidative damage being a non-redox metal but can subsequently lead to an increase in the Fe-induced ROS production and membrane peroxidation [82]. Humans are continuously exposed to multiple environmental metals and overexposure leads to neurotoxicity. More studies in this regard are necessary to find the synergetic effects of metals together and alone [79].

The brain microenvironment is separated from systemic circulation by choroid plexus (CP) and blood brain barrier (BBB) to defend and preserve the brain integrity from the effects of toxic metals and chemicals. Metals gains access to the human body by initially being absorbed and taken up from the gastrointestinal tract, through skin or lungs and then enter the systemic circulation [83]. The metals gain entry from blood to the brain and by crossing the CP and BBB, and then enter into the cerebrospinal fluid (CSF). From the CSF, entry is gained into specific regions of the brain. The BBB limits the entry of non-lipophilic substances inside and out of the brain. It has been reported that protective mechanisms like P- glycoprotein (MDR) and ATP binding cassette (ABCC) are present in the BBB to prevent the entry of harmful and toxic materials inside the brain [84].

Some neurotoxic metals like As, Pb, MeHg and Cd mimics the pathway of essential nutrients while crossing the BBB utilizing the ionic transporters [85]. Moreover, metals have an affinity for transferrin Tf-transporters and divalent metal ion transporter-I (DMT-I) that leads to interaction of metals in brain tissues and BBB. Studies have reported that Cd also have a very high permeability while crossing the BBB in rats. In vivo studies shows that Cd accumulates and penetrates in the adult and developing rat brains and to bind with metallothionein (MT-III) [86].

According to a study Cd, As and Pb produces a synergistic action by reducing the glial fibrillary acidic protein (GFAP) expression which is an important macromolecule for the BBB integrity [87]. It was also established by Rai *et al.*, that metal mixture (Cd, Pb, As) exposure leads to astrocyte toxicity resulting in damaging of the BBB and contributing to cognitive deficits [88]. Individuals are exposed to metal mixtures at different stages in life called as window of exposure *i.e.* embryo, fetus, child, adolescent, adult, old age. This exposure varies in different individuals depending on physiological and anatomical development of the BBB [84].

7. HARMFUL EFFECTS OF CHOLESTEROL IN ASSOCIATION WITH METAL EXPOSURE

A key risk factor for developing cognitive disorders is cholesterol and its different fractions. Lipids and lipoproteins play a vital role directly in forming amyloid plagues and hence result in neurodegenerative disorders [77]. These amyloid plagues cause cognitive disturbances by altering the synaptic plasticity, changes in tau phosphorylation and therefore end in neurological deficits [89]. The N-Methyl-D-aspartate (NMDA) receptors which are involved in long term potentiation in the hippocampus are impaired by triglycerides and results in elevated oxidative stress, resulting in cognitive impairment [90, 91]. HDL and LDL are reported to have contradictory effects on cognitive impairment. However, of both macromolecules, LDL cholesterol promotes the amyloid precursor protein internalization and disturbs the structure as well as function of neuronal lysosomes, thus increasing the amyloid beta accumulation [92]. However, HDL is found to have a significant anti-inflammatory and antioxidant potential [93].

Another study has found out that a significant link between the Cardiovascular and Cerebrovascular Diseases CCVD and higher levels of serum Cd and Pb. The chance of CCVD was found to be higher, still at minimal exposure of these metals. Not only this, but there was a strong correlation between the occurrence of CCVD and elevated levels of metal concentration like cadmium (Cd), tungsten (W), cobalt (Co) and antimony (Sb) [94]. Numerous studies are being carried out to create a link between clinical cardiovascular morbidity and Cd exposure, and one such finding suggests higher serum levels of Cd among people who had a myocardial infarction. Moreover, constant exposure to Pb and Cd suggest their plausible role in inflammation and impaired renal function which tend to exacerbate the pathogenesis of atherosclerosis and hence cause CCVD [94].

Another preliminary study by the American Heart Association's Scientific Session 2018 suggested that higher levels of lead (Pb) and other heavy metals were detected in the blood along with increased LDL levels. The researchers also reviewed the information form NHANES 2009-2012, a database which represents blood levels of heavy metals along with cholesterol levels [95]. The study found that there was a 56% chance of having higher LDL if the people had highest Pb levels, 73% more likely to have higher values of total cholesterol if they had highest Hg in the blood. The rise in the levels of cholesterol with increased metal levels might have cardiovascular consequences. Therefore, it is important to screen for heavy metals as a contributing factor for elevated cholesterol [95]. According to a study by Buhari et. al, a survey conducted from 2009-2012 National Health and Nutrition Examination Survey, it was found that there was a strong relationship of serum lead levels with LDL and total cholesterol. This may in turn contribute to cardiovascular diseases in population exposed to heavy metals [96].

A study suggests that there is evidence that higher total cholesterol levels may contribute to Alzheimer's disease (AD) and other related dementias in old aged individuals. It was found that total serum lipids were very high in patients who were already having some form of cognitive deficits, implying the role of cholesterol and its components in hastening the process of cognitive aging and eventually causing neurodegeneration [97]. In recent years, the presence of Apolipoprotein E4 (ApoE4) was found as a genetic risk factor for late AD onset. It is speculated that ApoE polymorphism accounts for abnormally higher levels of serum lipids, which contribute to early onset of coronary artery disease and atherosclerosis. It is suggested that defects in lipid metabolism due to dysfunctional ApoE4 may serve as a critical factor in AD development. Other conditions that may also arise due to ApoE polymorphism is the cognitive decline and AD [98]. However, the elevated serum lipid AD relationship is studied independent of ApoE 4 allele and was concluded that LDL and not HDL is correlated with AD pathology. Moreover, age is also an important risk factor in AD manifestation, the presence of ApoE4 in adults below the age of 60-65 is linked with higher LDL and higher total serum cholesterol levels [97].

How some toxic metals are associated with obesity or increased cholesterol levels, may partly because of the fact that these heavy metals induce oxidative stress which boosts lipogenesis and that in return increases the free radical generation. The ROS formed by metals interferes with the normal metabolic function of the mitochondria and inhibit energy production by oxidative phosphorylation. The lower levels of ATP produced along with altered tricarboxylic acid (TCA) efficacy due to inhibition of some enzymes sensitive to oxidative stress, cause the liver towards lipogenesis, away from the normal metabolism [99]. According to another study it was found that there was a strong correlation of blood and tibia Pb with renal function and diabetes with an implication that Pb exposure harms the kidneys more in diabetics as compared to the non-diabetics [100]. It has also been shown that acute and chronic Pb poisoning can cause cardiac issues and cerebrovascular diseases in Pb exposed workers compared with controls [101].

A study by Wojtczak-Jaroszowa and Kubow explains three possible mechanisms by which Pb can bring about pathophysiological disorders, especially atherosclerosis. Firstly, the inhibition of superoxide dismutase (SOD), that results in serum lipid peroxide elevation. Secondly, the formation of atherosclerotic plaques from a single proliferating mutant cell and lastly the inhibition of cytochrome P-450 activity, which contributes to subsequent increase in serum lipid levels. The major finding suggests that serum total lipoprotein and cholesterol levels were higher in patients who were continuously exposed to Pb as compared to non-exposed individuals. Moreover, it was found that there was a dose dependent response between serum cholesterol and blood Pb levels in people who were exposed, which suggests impaired metabolism of lipids due to exposure to Pb [102].

Furthermore, Pb and other heavy metals tend to decrease the activity of P-450, which limits the bile acids biosynthesis, leading to accumulation of cholesterol as it is the single important route for elimination of cholesterol from the body. In a study of Wister rats, it was found that increased synthesis of cholesterol may take place due to Pb stimulated increase in expression of hepatic enzymes for de novo cholesterol synthesis or due to altered feedback inhibition. Therefore, it was concluded that blood Pb levels are positively correlated with serum cholesterol, lipoprotein and HDL cholesterol levels [101].

Cd is also implicated in pathogenesis of several cardiovascular diseases like dyslipidemia and diabetes. It was found that oral administration of Cd led to higher levels of TG, TC and LDL in rats after three months of exposure. These results were in line with previous findings which suggested that Cd administration leads to an increase in lipid peroxidation. This contributes to several pathological processes due to increased oxidative stress [103]. Another study suggested that heavy metals levels in the blood are found to have an association with cardiovascular mortality and higher cholesterol levels. The presence of heavy metals not only disturbs the metabolic functions but also alter the cognitive performances [77]. An in-vitro study showed that the development of macrophage derived lipid cells are actively involved in oxidation of LDL in the presence of copper. High serum copper concentrations are found to be associated with atherosclerosis [104].

The hazardous and toxic nature of metals such a Cd, Hg and Pb may have many pathophysiological implications which include hypertension, coronary artery disease, cerebrovascular and renal dysfunctions. Moreover, these metals are involved in the formation of free radicals which lead to neurological deficits [96].

8. ROLE OF APOE GENOTYPE IN COGNITIVE IMPAIRMENT IN ASSOCIATION WITH METAL EXPOSURE

Apolipoproteins are the protein components of the lipoproteins like low density lipoprotein (LDL), very low density lipoprotein (VLDL), high density lipoprotein (HDL) etc. The apolipoproteins are present inside the molecular surface as well as transmembrane region. It is a glycoprotein of 34kDa and has various isoforms. ApoE is an important lipid metabolism lipoprotein in brain as well as rest of the body and is involved in transport and delivery of cholesterol to neurons. Brain is an important organ in the entire human body with highest levels of cholesterol, and apoE is the key apolipoprotein found in this region. Brain is the second largest and abundant site of apoE synthesis, produced majorly by astrocytes in order to transport the cholesterol to the neurons through receptors of apoE. The production of apoE is subject to neuronal injury or insult, in order to promote neuronal repair and integrity. Moreover, the synthesis of apoE is also essential for synaptic formation and axonal growth [105]. There are many studies regarding association of apoE4 and neurodegeneration in AD, and its expression is correlated with increased rates of tau hyper phosphorylation, facilitating the Neurofibrillary tangles (NFTs) formation. Moreover, it is also involved in production and deposition of Aβ, leading to the accumulation of senile plaque by impairing the Aβ clearance [106].

The apolipoprotein ApoE4 allele is present on chromosome 19 and it has been well documented that its presence in young carriers shows cognitive advancement as compared to the non-carriers. This renders ApoE a good candidate for genetic studies on cognition. Despite having this allele at an early age to be beneficial, ApoE is also known as a susceptibility gene for increasing the incidence of Alzheimer's disease (AD). Having inherited 1 or 2 ApoE4 alleles increase the risk of having AD but lowers the age of disease onset and decrease the performance of AD patients in various cognitive ability based test and also explains the 17% prevalence rate of AD in a population [107].

It is reported that changes in the overall size of the Golgi apparatus (GA) is used as a tool to measure proper neuronal metabolism. It is important because the proteins synthesized for fast axonal transport are routed through GA. In addition, it was established that AD patients having an ApoE4 allele had a reduced size of GA, providing evidence for decreased metabolism. Therefore, it was concluded that ApoE4 presence reduced the neuronal metabolism and therefore makes the neurons much more vulnerable to external insults and accelerates the process of aging along with AD [107]. According to Ghebremedhin *et al.*, a strong relation between the ApoE 4 allele presence and NFT formation at a very early stage of AD (Braak I) was established, at a young age, with a mean age of only 38 years in subjects [108].

• Evidence For The Link Between ApoE And Metals

According to a study old aged individuals have lower cerebral blood flow (CBF) as compared to younger individuals. It has been found that ApoE impacts cerebrovascular function and also influences the association between CBF and cognition [109]. ApoE4, a key genetic risk factor for AD, has also been implicated in small vessels disease (SVD), white matter lesions contributing to vascular cognitive impairment and Amyloid-β accumulation in AD. Therefore, in addition to reducing CBF, ApoE4 is also involved in the failure of basic homeostatic mechanisms of the brain regions, pertaining to cognitive deficits [110]. There are many studies pointing that Al exposure leads to a decline in cognitive functions and also cause neurodegenerative disorders, particularly AD. People who were exposed to Al contaminated water proved to have impaired memory and information processing was greatly altered. Moreover, a study reported the codeposition of Al with β-amyloid protein in the brain, providing basis for neurological impairments in the later life. There was also a connection between ApoE genotype and its impact on Al uptake, contributing to cognitive changes over lifetime [111].

The occurrence of ApoE4 allele and exposure to heavy metals together contribute to a decline in cognition. According to a study, the researchers established a mouse model with activated version of ApoE gene and then exposed the model to low levels of Cd in drinking water for a period of 14 weeks. They concluded that the cognitive abilities of mice were altered due to a dysfunctional hippocampus (a brain area important for memory and learning) and they performed poorly in standard novel object location and T-maze tests. This indicates a decline in poor short term spatial working memory. It was established that mice with ApoE4 gene performed worse in T maze test as compared to ApoE3. The authors suggest that there may be a possibility of ApoE4 causing leakage of the BBB and caused accumulation of Cd in ApoE4 brains, hence accelerating cognitive impairment and altered hippocampal neurogenesis [112].

According to recent studies, there has been an association between Hg intoxication and various isoforms of apolipoprotein E. It is proposed that ApoE4 tend to potentiate the damage caused by Hg and leads to an increase in altered motor coordination, paralysis and visual/tactile dysfunction as a consequence of neurodegeneration. Various diseases that may arise due to exposure to heavy metals along with ApoE4 polymorphism include amyotrophic lateral sclerosis, Alzheimer's disease and Parkinson's disease. The relationship of ApoE4 with the metals (Pb, Hg, Zn, Cu and Fe) seem to be toxicokinetic. *i.e.* the presence of different ApoE isoforms and its impact on clearance and bioavailability of the metals. It is of immense importance to note that individuals carrying one of two copies of ApoE4 allele are more susceptible to Hg intoxication [105].

The expression of ApoE4 and not ApoE2 and ApoE3 is involved with the breakdown of BBB by the proinflammatory cytokines expression like cyclophilin A and cause activation of matrix metalloproteinases and NF-κB [113]. In another study, it was found that older subjects carrying an ApoE4 showed higher percentage of BBB breakdown marker called csf/plasma albumin quotient values as compared to ApoE2/ApoE3 carriers. This change in the BBB integrity interferes with cognitive function [114].

Co exposure to metals like Hg along with being genetically predisposed to ApoE genotype leads to oxidative imbalance in the CNS, reduced levels of antioxidant enzymes, neuroinflammation, accumulation of amyloid beta, altered glutamate and γ-aminobutyric acid (GABA) signaling and tau hyper phosphorylation as shown in Fig. (**2**). Therefore, it is of immense importance to study the presence of ApoE4 allele and the toxicodynamic changes that are induced synergistically in the presence of mercury [105].

Mercury intoxication and Apo lipoprotein E.

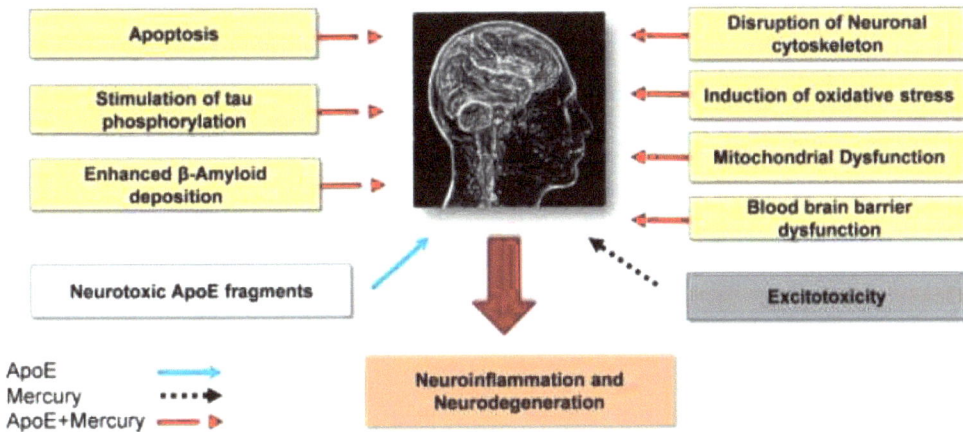

Fig. (2). The effect of Hg exposure on ApoE expression results in neurodegeneration. The black and blue arrows indicate mechanisms in the presence of only Hg and ApoE respectively. The red arrows show the mechanisms shared by both. It is speculated that the presence of ApoE mainly affects the toxicodynamic changes that may act synergistically with effects of Hg. There is a possibility that Hg and ApoE follow the same mechanisms of neurotoxicity leading to neurodegenerative illnesses. Adapted from [105].

Exposure to Pb along with the presence of ApoE genotype manifests long term neurophysiological insults like dementia. The ApoE4 induced lead neurotoxicity might surface either due to Pb exposure alone, or due to the presence of ApoE4 genotype or because of the combination of both ApoE genotype and Pb. Several mechanisms of actions are involved in Pb induced neurotoxicity, such as the involvement of triethyl lead in the inhibition of oxidative phosphorylation, phosphocreatine synthesis, amino acid metabolism etc. Moreover, the triethyl lead may also inhibit the synthesis of microtubules leading to the shrinking and clumping of cells, eventually causing cellular necrosis and vacuolar degeneration [115].

In addition to the association of ApoE4 with metals to exacerbate cognitive dysfunction, it is also reported that the presence of ApoE4, disrupts the function of α-synuclein. ApoE4 leads to a negative impact on lipid metabolism and synaptic function because both ApoE4 and α-synuclein bind to the lipids, it is probable that they compete for membrane lipid binding and thus have neurological consequences [116]. The ApoE4 allele is a key player in neurite growth and regulates synaptogenesis, along with its ability to scavenge cholesterol for regeneration of neurons. It was demonstrated that ApoE deficient mice had cognitive dysfunctions and exhibited memory deficits in the Morris water maze test [117].

The neurological deficits due to the presence of ApoE genotype are also found to be triggered by imbalance of some other metals like Zn and Cu. The amount of free Cu in blood was found to be elevated in AD as compared to the controls. Moreover, the levels of Fe and Zn were found to be elevated in brain parenchyma in AD subjects. Zn as well as Fe and Cu which are released during the neural transmission directly attach to Aβ and lead to accumulation and aggregation into the amyloid plagues [118]. The ApoE isoforms tend to bind to metals like Zn, Fe and Cu, thus contributing to AD pathogenesis. There is a key role of ApoE in synaptic glutamate and Zn levels regulation in the regions of hippocampus. Upon deletion of ApoE, there is a reduction in Zn levels, suggesting that ApoE somehow control Zn levels in the brain. The amount of Zn required in the synapses for long term potentiation serves as an important factor for proper functioning of hippocampus. Therefore, decreased ApoE levels, results in decreased synaptic Zn and thus lead to cognitive impairments as shown in Fig. (**3**). Taken these findings, it is evident that there is a strong interplay between metals and ApoE and they may work in concert in AD pathogenesis [119].

Mechanism of action of apoE and metals in AD associated pathogenesis.

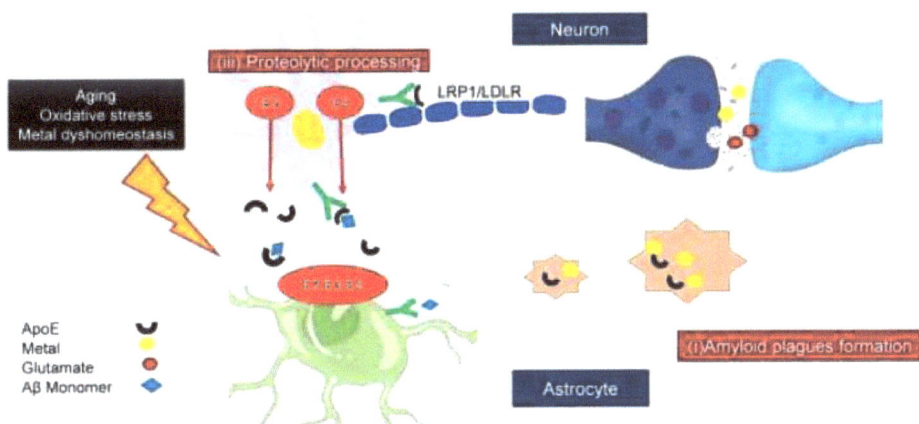

Fig. (3). Mechanism of action of ApoE and metals in AD associated pathogenesis. A number of mechanisms are involved in metal and ApoE interaction. (i) Metals like Cu, Fe, Zn, Al, along with ApoE results in amyloid- β aggregation and cause metal dyshomeostasis and as consequence the ApoE levels are decreased in brain. (ii) Such metal dyshomeostasis in AD subjects influence the ApoE expression levels in astrocytes, this happens either due to low transcriptional activity levels in the nucleus or ER where ApoE is synthesized. ApoE mediates the removal and clearance of Aβ through the LRP1/LDLR (Low density lipoprotein receptor-related protein 1), therefore, reduced ApoE levels cause an increase in aggregation of Aβ. (iii) As a result of amyloid formation and increase in oxidative stress, the ApoE expression in increased. However, neuronal ApoE is processed and fragments are generated that are truncated. In comparison with ApoE2 and ApoE3, ApoE4 is more susceptible to degradation. During the ApoE proteolysis, metals bind to ApoE2 and ApoE4 is degraded. These ApoE4 fragments induce cytoskeletal and mitochondrial abnormalities and consistent reduction in ApoE4 expression cause a damage to cognitive abilities by hippocampal LTP functioning. Adapted from (Xu *et al.*, 2014).

According to studies, ApoE plays a crucial role in manifestation of cognitive disorders based on the genotype. Those who were homozygous for ApoE allele were more susceptible to cognitive impairments in comparison to the heterozygous individuals [120]. ApoE4 is multifunctional and its targeted knockout leads to loss of synaptic plasticity associated with memory and learning, NFT formation, Aβ oligomer production, increased oxidative stress and ultimately neuronal death [121]. There is another important aspect to analyze the association between ApoE presence and exposure to environmental toxins, especially metals. Such studies focusing on gene and environment together, direct us towards better understanding of cognitive impairments. For instance, it was reported both through epidemiological and animal studies that exposure to lead results in cognitive malfunctioning. Moreover, those workers who were occupationally exposed to Pb with presence of at least one ApoE4 allele experienced a more progressive decline in cognitive functions as compared to ApoE4 non carriers [122].

9. ROLE OF TOTAL TAU IN COGNITIVE IMPAIRMENT (IN ASSOCIATION WITH METAL EXPOSURE)

Multiple studies have reported the presence of total tau in patients with mild cognitive impairment (MCI) and AD. A recent study reported that increased total tau levels were related to an increased ventricular volume and was major factor of cognitive decline in MCI, AD and dementia subjects. Moreover, it was also noted that there was an increased interaction between plasma total tau and Aβ levels in people with cognitive decline, and this was an important factor in MCI progression [123]. Tau protein is present in the axonal regions of neuronal cells and is important for the dynamics and functioning of microtubules. When the neurons are damaged, it is released into the peripheral blood as well as cerebrospinal fluid (CSF).

Exposure to certain environmental toxins especially heavy metals such as Al and Pb are linked with AD pathology. People working in the industries involved in the foundry process are continuously exposed to metals like Al, Mn, Pb and Zn. There have been many studies regarding Al neurotoxicity and rest of the metals follow the same mechanism of action in contributing towards cognitive impairments [124]. Al neurotoxicity is linked to tau dependent mechanisms, where large concentrations of neurofibrillary tangles were observed in the brain regions with higher Al concentrations. Moreover, it was also suggested that elevated levels of Al, facilitated tau protein hyperphosphorylation and aggregation resulting in tangle formation [125]. Co-exposure of other metals was also reviewed, and the results showed that there were higher levels of Mn and Pb with lower Zn levels in the foundry workers in comparison to the normal individuals. Very low serum concentration of Mn have been previously related to neurobehavioral deficits and cognitive decline [126]. Phosphorylated tau has been a reliable diagnostic tool for mild cognitive impairment and for MCI progression. Al has been found to play an active role in tau hyperphosphorylation, the phosphorylated tau may be used as an essential indicator to evaluate the deterioration of cognitive function, and therefore, p-tau 181 and p-tau 231 were used in a study for cognitive decline prediction. It was concluded the lymphocytes of Al exposed workers had high levels of p-tau 181 and p-tau 231 [127].

10. ROLE OF AMYLOID B-42 IN COGNITIVE IMPAIRMENT (IN ASSOCIATION WITH METAL EXPOSURE)

Along with aging, the exposure to heavy metals potentially induces neurodegenerative alterations. Despite the importance of these metals physiologically in different intracellular processes, their excessive amounts contribute to neurotoxicity. Clinical findings suggest abnormally higher levels of Zn, Fe and Cu in post mortem AD brain tissues. Moreover, Al, Pb, Li, Cd and Hg exposure lead to AD associated neurological pathologies. Al has also been detected in Aβ plagues and NFT's, the exposure of neuronal cultures to Al results in marked Aβ accumulation *in vitro*. Along with the involvement of Al in Aβ pathology, Al promotes tau accumulation and aggregation by the PP2A reduction along with an increase in CDK5 kinase levels and GSK3- β kinase activity as shown in Fig. (**4**) [128].

Effect of metal ions on tau and amyloid β (Aβ) aggregation.

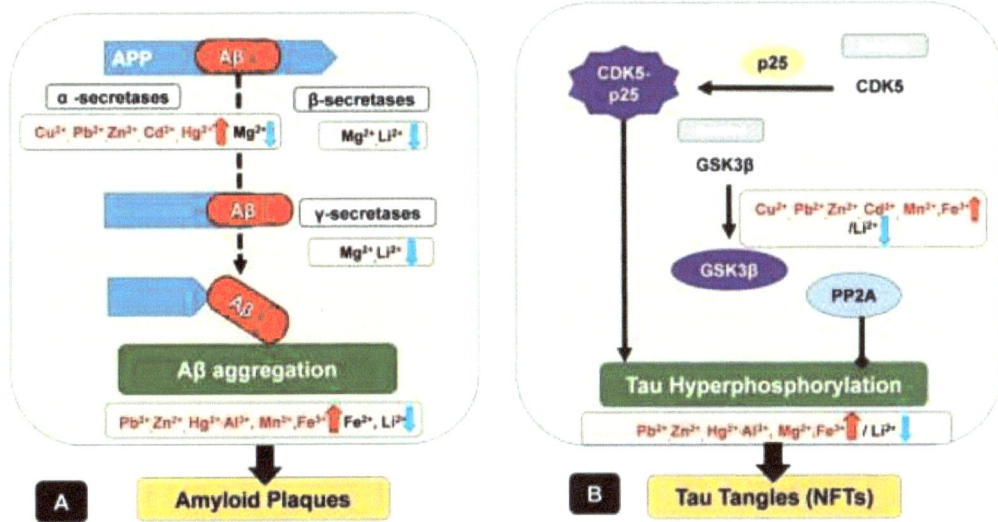

Fig. (**4**). (**a**) **Amyloidogenesis**. During normal circumstances, Aβ is rarely formed in the brain. In pathological conditions, the APP is cleaved by β- and γ-secretases, leading to Aβ neurotoxic peptides formation. This Aβ is released into the extracellular space transforming into the amyloid plaque fibrils. Metals like Fe^{3+}, Cu^{2+}, Pb^{2+}, Mn^{2+}, Zn^{2+}, Cd^{2+}, Al^{3+} and Hg^{2+} induce Aβ aggregation (red arrows). In comparison, Fe^{2+}, Li^{2+} and Mg^{2+} reduce Aβ formation (blue arrows). (**b**) **Tau pathology**. In order to maintain neuronal structure and function the tau protein is constantly phosphorylated and dephosphorylated. But during pathological condition, tau phosphorylation is elevated by cyclin dependent kinase 5 (CDK5) and glycogen synthase kinase 3-beta (GSK-3). This condition of tau hyper phosphorylation is further maintained by inactivity of phosphatase like protein phosphatase 2A (PP2A). This hyper phosphorylation leads to neurofibrillary tangles (NFTs) aggregation. Adapted from [128].

Progressive supranuclear palsy (PSP) is a neurodegenerative condition associated with overall reduction of different brain regions, which progresses into dementia. Along with other genes responsible for the manifestation of this condition, the MAPT gene, which encodes tau protein is largely related with PSP. Exposure to various heavy metals like Ni, Cr, and Cd from improper disposal of waste from industries elevates the risk of sporadic PSP. According to a study, treatment of induced pleuripotent stem cells (having a PSP-mutation in MAPT gene) with these metals caused cell death in a dose dependent manner. Furthermore, Ni and Cr treatment to SH-SY5Y cells (which is a well-established dopaminergic neuronal cell model) induced apoptosis and cell death. These findings suggest that heavy metals exposure in connection with increased tau protein formation is linked to neurodegenerative disorders like PSP [129].

Abnormal flux of metal ions into the brain through the BBB leads to their accumulation in brain. In this regard the metal transporters are key elements in their uptake and accumulation. Metals sharing the same type of transporter, tend to compete with one another, for example transferrin (Tf). Tf binds to several metals like Cd, Co, Mn, Zn, Cr, Al and Fe, and is found in large concentrations in senile plaques due to its binding to Zn and Fe [130]. Al, Fe and Zn also accelerate the process of accumulation of Aβ *in vitro*. Once the Aβ aggregates are generated, even at very low amyloid levels helps in its growth and progression. Therefore, by promoting the initial Aβ aggregation, Al contributes to the accumulation of amyloid either alone or in combination with other ligands. Al salts, at concentration of greater than or equal to 100 uM induce tau aggregation that prevents the entry into the SDS- polyacrylamide gels. Therefore, excess exposure to Aluminium results in cognitive deficits, peripheral neuropathy and disorders of motor control [131].

Aβ is formed by the cleavage of the amyloid precursor protein by alpha and beta secretases. The most important isoforms of Aβ are Aβ40 and Aβ42. Of these isoforms, Aβ42 is less abundant but more neurotoxic. Specific metals are observed in lesions of the AD autopsy patients. It showed abnormal levels of Cu, Zn, Fe, whereas Al was also found in amyloid fibers in the core of senile plaque. It was further concluded that Zn, without undergoing a nucleation phase, facilitates the annular protofibril formation leading to an abnormal amount of Aβ deposition [132]. The accumulated Aβ aggregate and form a pore like channel on the membranes [133].

The Aβ channels are unregulated and therefore, Ca^{2+} flows freely, leading to a number of neurodegenerative triggers like ROS formation, tau phosphorylation and disturbance of Ca^{2+} levels in the ER. The increase in the levels of Ca^{2+} can consequently result in the APP overproduction leading to increased Aβ levels, and ultimately an unregulated vicious cycle of Ca^{2+} influx is generated. It is possible that metals like Cu, Zn, Mn and Al accelerate the oligomerization of Aβ and therefore increase the rate of Aβ channel formation as shown in Fig. (**5**) . Moreover, the size and shape of Aβ oligomers formed in the presence of different metals are unique. It was also concluded that the Zn aggregated Aβ were comparatively less toxic to Cu aggregated Aβ. The effect of Zn is also observed in type 2 diabetes mellitus, which is secreted along with insulin from pancreatic-β cells, proves to have a dual impact on cell toxicity and Aβ aggregation. In pathological conditions, overexpression of genetic APP accumulate Aβ peptides, and incorporate directly into the membranes to form many amyloid channels [133].

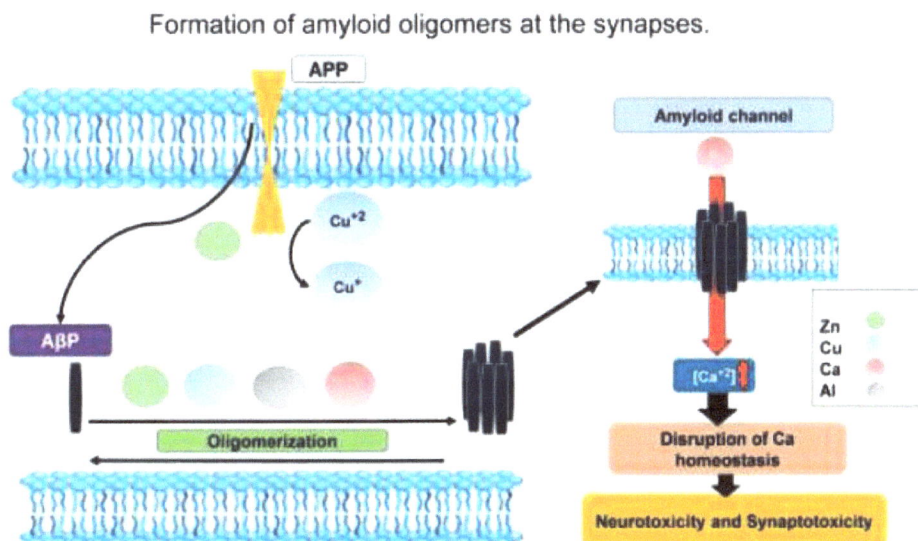

Fig. (5). hypothesis regarding toxic functions of the amyloid oligomers at the neuronal synapse. Aβ peptides are formed from the APP into synaptic clefts, and are then degraded by different proteases. However, elevated amounts of AβP or a different ratio of Aβ42 to Aβ40 caused due to the mutation of presenilins or APP may lead to an increase in AβP half-life and concentration in brain. The exposure to Al and other metals leads to the upregulation of APP and consequently increased AβP amounts. These AβP aggregate and results in the formation of pore like channels. As AβP channels are not regulated, Ca^{2+} freely flows through them and triggers a cascade of neuroinflammatory responses like ROS generation, altered Ca^{2+} levels and tau hyperphosphorylation. Elevated levels of Ca^{2+} results in overexpression of APP and increased AβP formation, leading to an unregulated vicious cycle of Ca^{2+} influx. Adapted from [133].

CONCLUDING REMARKS

Exposure and bioaccumulation of various heavy metals have increased exponentially in recent years, mainly due to rapid industrialization and growth of the human population. While such industrializations have improved the quality of life, on the other hand it has contributed immensely to health hazards. These toxic metals discharged into the environment have become part of our food chain. Studies suggest that constant exposure to even low levels of these environmental pollutants cause sustained damage to the brain.

The present studies point towards Al being an exacerbating factor for dementia and AD, rather than being the core reason of the disease. Use of Al salts dates back to the Roman times, where it was implied for the treatment of wounds, purification of water and gastrointestinal issues. One of the earliest findings suggested that the point of attack of Al is the CNS, and specifically elevated levels were found in the neocortical regions of people affected by AD. Not only Al but the imbalance of other metal ions triggers the protein aggregation and misfolding, mitochondrial dysfunction and cause neurodegenerative diseases. Transition metals like Cu, Hg and Mn has strong neurotoxic potential. Heavy metals, for example Cd, As and Pb bind to the proteins and interfere with protein function and metal transport.

ApoE is a serum transport protein for fatty acids and cholesterol found in the cytoplasm, which binds to the LDL and ApoE specific receptors. It is also one of the major lipoprotein produced in the brain parenchyma and controls the normal neural transmission. The interaction of ApoE or cholesterol with metals leads to the aggregation of Aβ, contributing to neuritic plaques and ultimately results in cognitive deficits. It is evident that the ratio of phospholipids to cholesterol is affected by different genetic (ApoE) and environmental factors and contributes to AD pathogenesis. Cholesterol tends to alter the fluidity of the membrane and cause oligomerization of the proteins, therefore, having an impact on channel formation which disturbs the BBB and facilitate the entry of neurotoxins. Therefore, it can be concluded that such neurodegenerative diseases are not only genetic or environmental in origin but a result of gene and environment together. Moreover, metals act as cofactors for many proteins, it is important to consider that higher levels of these metals aggravate the disease progression by facilitating tau and Aβ production as well as an increase in amyloid precursor protein APP processing.

Despite the fact that major attention is focused on public health in terms of toxicants causing neurological illnesses, there is still a constant increase in the environmental pollutants. In the presence of adequate measures and proper scrutiny before the release of these toxic metals into the environment, it is the need of hour to have sufficient knowledge about long term implications that such toxic metals may have on individual or health of masses. In a time of unprecedented chemical/metal exposure as confirmed by medical organizations such as Centre for Disease Control, it is suggested that people having cognitive impairment related issues and dementia should be investigated for exposure to such toxicants. These investigations will have a positive outcome on our current and future generations in terms of determining the risk factors.

CONSENT FOR PUBLICATION

Not Applicable.

CONFLICT OF INTEREST

The author declares no conflict of interest, financial or otherwise.

ACKNOWLEDGEMENTS

Declared none.

LIST OF ABBREVIATIONS

AD Alzheimer's Disease

PD Parkinson's Disease

APP Amyloid Precursor Protein

ApoE Apolipoprotein E

Aβ Amyloid Beta

AβP Amyloid Beta peptides

ROS Reactive Oxygen species

RNS Reactive Nitrogen species

LPO Lipid peroxidation

PVC Polyvinyl chloride

HFD High Fat Diet

iNOS Inducible nitric oxide synthase

TNF-α Tumour Necrosis Factor alpha

NF-κB Nuclear Factor kappa-light-chain-enhancer of activated B cells

MAPKs Mitogen activated protein kinases

BBB Blood brain barrier

CSF Cerebrospinal fluid

GFAP Glial fibrillary acidic protein

DMT-I Divalent metal ion Transporter-I

NMDA N-Methyl-D-aspartate

CCVD Cardiovascular and Cerebrovascular Diseases

LDL Low density lipoprotein

HDL High density lipoprotein

TG Triglycerides

TC Total cholesterol

CBF Cerebral blood flow

NFTs Neurofibrillary tangles formation.

MCI Mild cognitive impairment

GSK3- β Glycogen synthase kinase 3 beta

PP2A Phosphatase like protein phosphatase 2A

CDK5 Cyclin dependent kinase 5

PSP Progressive supranuclear palsy

Fe Iron

Zn Zinc

Cu Copper

Hg Mercury

Al Aluminium

Cd Cadmium

Ni Nickle

Co Cobalt

Pb Lead

As Arsenic

Mn Manganese

REFERENCES

[1]   Giagulli VA, Guastamacchia E, Licchelli B, Triggiani V. A Giagulli V, Guastamacchia E, Licchelli B, Triggiani V. Serum testosterone and cognitive function in ageing male: updating the evidence. Recent Pat Endocr Metab Immune Drug Discov 2016; 10(1): 22-30.
[http://dx.doi.org/10.2174/1872214810999160603213743] [PMID: 27981914]

[2]   Plassman BL, Langa KM, Fisher GG, *et al.* Prevalence of cognitive impairment without dementia in the United States. Ann Intern Med 2008; 148(6): 427-34.
[http://dx.doi.org/10.7326/0003-4819-148-6-200803180-00005] [PMID: 18347351]

[3]   Deary IJ, Corley J, Gow AJ, *et al.* Age-associated cognitive decline. Br Med Bull 2009; 92(1): 135-52.
[http://dx.doi.org/10.1093/bmb/ldp033] [PMID: 19776035]

[4]   Mattson MP, Magnus T. Ageing and neuronal vulnerability. Nat Rev Neurosci 2006; 7(4): 278-94.
[http://dx.doi.org/10.1038/nrn1886] [PMID: 16552414]

[5]   Harper ME, Bevilacqua L, Hagopian K, Weindruch R, Ramsey JJ. Ageing, oxidative stress, and mitochondrial uncoupling. Acta Physiol Scand 2004; 182(4): 321-31.
[PMID: 15569093]

[6]   Meramat A, Rajab NF, Shahar S, Sharif RA. DNA damage, copper and lead associates with cognitive function among older adults. J Nutr Health Aging 2017; 21(5): 539-45.
[http://dx.doi.org/10.1007/s12603-016-0759-1] [PMID: 28448084]

[7]   Afridi HI, Kazi TG, Kazi AG, *et al.* Levels of arsenic, cadmium, lead, manganese and zinc in biological samples of paralysed steel mill workers with related to controls. Biol Trace Elem Res 2011;

144(1-3): 164-82.
[http://dx.doi.org/10.1007/s12011-011-9063-4] [PMID: 21547399]

[8]   Jomova K, Valko M. Advances in metal-induced oxidative stress and human disease. Toxicology 2011; 283(2-3): 65-87.
[http://dx.doi.org/10.1016/j.tox.2011.03.001] [PMID: 21414382]

[9]   Iqbal G, Zada W, Mannan A, Ahmed T. Elevated heavy metals levels in cognitively impaired patients from Pakistan. Environ Toxicol Pharmacol 2018; 60: 100-9.
[http://dx.doi.org/10.1016/j.etap.2018.04.011] [PMID: 29684799]

[10]   Wong WY, Flik G, Groenen PMW, *et al.* The impact of calcium, magnesium, zinc, and copper in blood and seminal plasma on semen parameters in men. Reprod Toxicol 2001; 15(2): 131-6.
[http://dx.doi.org/10.1016/S0890-6238(01)00113-7] [PMID: 11297872]

[11]   Baxter JH, Van Wyk JJ, Follis RH Jr. A bone disorder associated with copper deficiency. II. Histological and chemical studies on the bones. Bull Johns Hopkins Hosp 1953; 93(1): 25-39.
[PMID: 13082312]

[12]   Li Y, Togashi Y, Sato S, *et al.* Spontaneous hepatic copper accumulation in Long-Evans Cinnamon rats with hereditary hepatitis. A model of Wilson's disease. J Clin Invest 1991; 87(5): 1858-61.
[http://dx.doi.org/10.1172/JCI115208] [PMID: 2022751]

[13]   Kok FJ, Van Duijn CM, Hofman A, *et al.* Serum copper and zinc and the risk of death from cancer and cardiovascular disease. Am J Epidemiol 1988; 128(2): 352-9.
[http://dx.doi.org/10.1093/oxfordjournals.aje.a114975] [PMID: 3394701]

[14]   Weihl CC, Lopate G. Motor neuron disease associated with copper deficiency. Muscle Nerve 2006; 34(6): 789-93.
[http://dx.doi.org/10.1002/mus.20631] [PMID: 16929546]

[15]   Syme CD, Nadal RC, Rigby SE, Viles JH. Copper binding to the amyloid-β (Abeta) peptide associated with Alzheimer's disease: folding, coordination geometry, pH dependence, stoichiometry, and affinity of Abeta-(1-28): insights from a range of complementary spectroscopic techniques. J Biol Chem 2004; 279(18): 18169-77.
[http://dx.doi.org/10.1074/jbc.M313572200] [PMID: 14978032]

[16]   Keith S, Jones D, Rosemond Z, Ingerman L, Chappell L. 2008.

[17]   Costello S, Brown DM, Noth EM, *et al.* Incident ischemic heart disease and recent occupational exposure to particulate matter in an Aluminium cohort. J Expo Sci Environ Epidemiol 2014; 24(1): 82-8.
[http://dx.doi.org/10.1038/jes.2013.47] [PMID: 23982120]

[18]   Kurella M, Chertow GM, Luan J, Yaffe K. Cognitive impairment in chronic kidney disease. J Am Geriatr Soc 2004; 52(11): 1863-9.
[http://dx.doi.org/10.1111/j.1532-5415.2004.52508.x] [PMID: 15507063]

[19]   Mazzoli-Rocha F, Dos Santos AN, Fernandes S, *et al.* Pulmonary function and histological impairment in mice after acute exposure to Aluminium dust. Inhal Toxicol 2010; 22(10): 861-7.
[http://dx.doi.org/10.3109/08958378.2010.489074] [PMID: 20545475]

[20]   Crapper McLachlan D. Aluminium and Alzheimer's disease. Neurobiol Aging 1986.
[http://dx.doi.org/10.1016/0197-4580(86)90102-8]

[21]   Hirsch EC, Brandel JP, Galle P, Javoy-Agid F, Agid Y. Iron and Aluminium increase in the substantia nigra of patients with Parkinson's disease: an X-ray microanalysis. J Neurochem 1991; 56(2): 446-51.
[http://dx.doi.org/10.1111/j.1471-4159.1991.tb08170.x] [PMID: 1988548]

[22]   Larson EB, Kukull WA, Katzman RL. Cognitive impairment: dementia and Alzheimer's disease. Annu Rev Public Health 1992; 13(1): 431-49.
[http://dx.doi.org/10.1146/annurev.pu.13.050192.002243] [PMID: 1599598]

[23] Galdes A, Vallee BL. Categories of zinc metalloenzymes. Met Ions Biol Syst 1983; 15: 1-54.

[24] Fine JM, Gordon T, Chen LC, Kinney P, Falcone G, Beckett WS. Metal fume fever: characterization of clinical and plasma IL-6 responses in controlled human exposures to zinc oxide fume at and below the threshold limit value. J Occup Environ Med 1997; 39(8): 722-6.
[http://dx.doi.org/10.1097/00043764-199708000-00006] [PMID: 9273875]

[25] Cuajungco MP, Fagét KY. Zinc takes the center stage: its paradoxical role in Alzheimer's disease. Brain Res Brain Res Rev 2003; 41(1): 44-56.
[http://dx.doi.org/10.1016/S0165-0173(02)00219-9] [PMID: 12505647]

[26] Lidsky TI, Schneider JS. Lead neurotoxicity in children: basic mechanisms and clinical correlates. Brain 2003; 126(Pt 1): 5-19.
[http://dx.doi.org/10.1093/brain/awg014] [PMID: 12477693]

[27] Needleman HL, Gatsonis CA. Low-level lead exposure and the IQ of children. A meta-analysis of modern studies. JAMA 1990; 263(5): 673-8.
[http://dx.doi.org/10.1001/jama.1990.03440050067035] [PMID: 2136923]

[28] Bouchard MF, Sauvé S, Barbeau B, et al. Intellectual impairment in school-age children exposed to manganese from drinking water. Environ Health Perspect 2011; 119(1): 138-43.
[http://dx.doi.org/10.1289/ehp.1002321] [PMID: 20855239]

[29] Finley JW. Does environmental exposure to manganese pose a health risk to healthy adults? Nutr Rev 2004; 62(4): 148-53.
[http://dx.doi.org/10.1301/nr.2004.apr.148-153] [PMID: 15141430]

[30] Lozoff B, Jimenez E, Hagen J, Mollen E, Wolf AW. Poorer behavioral and developmental outcome more than 10 years after treatment for iron deficiency in infancy. Pediatrics 2000; 105(4): e51-.
[http://dx.doi.org/10.1542/peds.105.4.e51]

[31] Kazi TG, Afridi HI, Kazi N, et al. Copper, chromium, manganese, iron, nickel, and zinc levels in biological samples of diabetes mellitus patients. Biol Trace Elem Res 2008; 122(1): 1-18.
[http://dx.doi.org/10.1007/s12011-007-8062-y] [PMID: 18193174]

[32] Powers KM, Smith-Weller T, Franklin GM, Longstreth WT Jr, Swanson PD, Checkoway H. Parkinson's disease risks associated with dietary iron, manganese, and other nutrient intakes. Neurology 2003; 60(11): 1761-6.
[http://dx.doi.org/10.1212/01.WNL.0000068021.13945.7F] [PMID: 12796527]

[33] Mergler D, Baldwin M, Bélanger S, et al. Manganese neurotoxicity, a continuum of dysfunction: results from a community based study. Neurotoxicology 1999; 20(2-3): 327-42.
[PMID: 10385894]

[34] Orłowski C, Piotrowski JK. Biological levels of cadmium and zinc in the small intestine of non-occupationally exposed human subjects. Hum Exp Toxicol 2003; 22(2): 57-63.
[http://dx.doi.org/10.1191/0960327103ht326oa] [PMID: 12693828]

[35] Nordberg GF. Cadmium and health in the 21st century--historical remarks and trends for the future. Biometals 2004; 17(5): 485-9.
[http://dx.doi.org/10.1023/B:BIOM.0000045726.75367.85] [PMID: 15688851]

[36] Kazantzis G. Renal tubular dysfunction and abnormalities of calcium metabolism in cadmium workers. Environ Health Perspect 1979; 28: 155-9.
[http://dx.doi.org/10.1289/ehp.7928155] [PMID: 488032]

[37] Waalkes MP, Rehm S, Riggs CW, et al. Cadmium carcinogenesis in male Wistar [Crl:(WI)BR] rats: dose-response analysis of tumor induction in the prostate and testes and at the injection site. Cancer Res 1988; 48(16): 4656-63.
[PMID: 3396014]

[38] Bar-Sela S, Reingold S, Richter ED. Amyotrophic lateral sclerosis in a battery-factory worker exposed

to cadmium. Int J Occup Environ Health 2001; 7(2): 109-12.
[http://dx.doi.org/10.1179/oeh.2001.7.2.109] [PMID: 11373040]

[39]   Popkin BM. Symposium: Obesity in developing countries: Biological and ecological factors. J Nutr 2001; 131(3)
[PMID: 11238777]

[40]   Drewnowski A, Popkin BM. The nutrition transition: new trends in the global diet. Nutr Rev 1997; 55(2): 31-43.
[http://dx.doi.org/10.1111/j.1753-4887.1997.tb01593.x] [PMID: 9155216]

[41]   Sellbom KS, Gunstad J. Cognitive function and decline in obesity. J Alzheimers Dis 2012; 30(s2) (Suppl. 2): S89-95.
[http://dx.doi.org/10.3233/JAD-2011-111073] [PMID: 22258511]

[42]   Pistell PJ, Morrison CD, Gupta S, *et al.* Cognitive impairment following high fat diet consumption is associated with brain inflammation. J Neuroimmunol 2010; 219(1-2): 25-32.
[http://dx.doi.org/10.1016/j.jneuroim.2009.11.010] [PMID: 20004026]

[43]   Morrison CD, Pistell PJ, Ingram DK, *et al.* High fat diet increases hippocampal oxidative stress and cognitive impairment in aged mice: implications for decreased Nrf2 signaling. J Neurochem 2010; 114(6): 1581-9.
[http://dx.doi.org/10.1111/j.1471-4159.2010.06865.x] [PMID: 20557430]

[44]   Kosari S, Badoer E, Nguyen JC, Killcross AS, Jenkins TA. Effect of western and high fat diets on memory and cholinergic measures in the rat. Behav Brain Res 2012; 235(1): 98-103.
[http://dx.doi.org/10.1016/j.bbr.2012.07.017] [PMID: 22820146]

[45]   Sharma DR, Wani WY, Sunkaria A, *et al.* Quercetin protects against chronic Aluminium-induced oxidative stress and ensuing biochemical, cholinergic, and neurobehavioral impairments in rats. Neurotox Res 2013; 23(4): 336-57.
[PMID: 22918785]

[46]   Batool Z, Sadir S, Liaquat L, *et al.* Repeated administration of almonds increases brain acetylcholine levels and enhances memory function in healthy rats while attenuates memory deficits in animal model of amnesia. Brain Res Bull 2016; 120: 63-74.
[http://dx.doi.org/10.1016/j.brainresbull.2015.11.001] [PMID: 26548495]

[47]   Gallagher M, Pelleymounter MA. Spatial learning deficits in old rats: a model for memory decline in the aged. Neurobiol Aging 1988; 9(5-6): 549-56.
[http://dx.doi.org/10.1016/S0197-4580(88)80112-X] [PMID: 3062465]

[48]   Thomas VS, Darvesh S, MacKnight C, Rockwood K. Estimating the prevalence of dementia in elderly people: a comparison of the Canadian Study of Health and Aging and National Population Health Survey approaches. Int Psychogeriatr 2001; 13(S1) (Suppl. 1): 169-75.
[http://dx.doi.org/10.1017/S1041610202008116] [PMID: 11892964]

[49]   Mohan R. Fiscal challenges of population aging: The Asian experience. Reserve Bank of India Bulletin, 2004.

[50]   2016 Alzheimer's disease facts and figures. Alzheimers Dement 2016; 12(4): 459-509.
[http://dx.doi.org/10.1016/j.jalz.2016.03.001] [PMID: 27570871]

[51]   Ritchie K. Mild cognitive impairment: an epidemiological perspective. Dialogues Clin Neurosci 2004; 6(4): 401-8.
[http://dx.doi.org/10.31887/DCNS.2004.6.4/kritchie] [PMID: 22034212]

[52]   Gauthier S, Reisberg B, Zaudig M, *et al.* Mild cognitive impairment. Lancet 2006; 367(9518): 1262-70.
[http://dx.doi.org/10.1016/S0140-6736(06)68542-5] [PMID: 16631882]

[53]   Azizullah A, Khattak MNK, Richter P, Häder D-P. Water pollution in Pakistan and its impact on public health--a review. Environ Int 2011; 37(2): 479-97.

[http://dx.doi.org/10.1016/j.envint.2010.10.007] [PMID: 21087795]

[54]   Nasrullah RN, Bibi H, Iqbal M, Durrani MI. Pollution load in industrial effluent and ground water of Gadoon Amazai Industrial Estate (GAIE) Swabi, NWFP. J Agric Biol Sci 2006; 1(3): 18-24.

[55]   Olness K. Effects on brain development leading to cognitive impairment: a worldwide epidemic. J Dev Behav Pediatr 2003; 24(2): 120-30.
        [http://dx.doi.org/10.1097/00004703-200304000-00009] [PMID: 12692458]

[56]   Ochmański W, Barabasz W. [Aluminium--occurrence and toxicity for organisms]. Przegl Lek 2000; 57(11): 665-8.
        [PMID: 11293216]

[57]   Wills MR, Hewitt CD, Sturgill BC, Savory J, Herman MM. Long-term oral or intravenous Aluminium administration in rabbits. I. Renal and hepatic changes. Ann Clin Lab Sci 1993; 23(1): 1-16.
        [PMID: 7679266]

[58]   Kumar V, Gill KD. Aluminium neurotoxicity: neurobehavioural and oxidative aspects. Arch Toxicol 2009; 83(11): 965-78.
        [http://dx.doi.org/10.1007/s00204-009-0455-6] [PMID: 19568732]

[59]   Exley C. The pro-oxidant activity of Aluminium. Free Radic Biol Med 2004; 36(3): 380-7.
        [http://dx.doi.org/10.1016/j.freeradbiomed.2003.11.017] [PMID: 15036357]

[60]   Platt B, Fiddler G, Riedel G, Henderson Z. Aluminium toxicity in the rat brain: histochemical and immunocytochemical evidence. Brain Res Bull 2001; 55(2): 257-67.
        [http://dx.doi.org/10.1016/S0361-9230(01)00511-1] [PMID: 11470325]

[61]   Kumar A, Singh A, Dogra S. POSSIBLE NEUROPROTECTIVE MECHANISM OF ANTIOXIDANTS AGAINST ALUMINIUM INDUCED COGNITIVE DYSFUNCTION.

[62]   Kumar V, Bal A, Gill KD. Impairment of mitochondrial energy metabolism in different regions of rat brain following chronic exposure to aluminium. Brain Res 2008; 1232: 94-103.
        [http://dx.doi.org/10.1016/j.brainres.2008.07.028] [PMID: 18691561]

[63]   Leuner B, Glasper ER, Mirescu C. A critical time for new neurons in the adult hippocampus. J Neurosci 2007; 27(22): 5845-6.
        [http://dx.doi.org/10.1523/JNEUROSCI.1838-07.2007] [PMID: 17537953]

[64]   Hirai K, Aliev G, Nunomura A, *et al.* Mitochondrial abnormalities in Alzheimer's disease. J Neurosci 2001; 21(9): 3017-23.
        [http://dx.doi.org/10.1523/JNEUROSCI.21-09-03017.2001] [PMID: 11312286]

[65]   Harrington CR, Wischik CM, McArthur FK, Taylor GA, Edwardson JA, Candy JM. Alzheimer's-disease-like changes in tau protein processing: association with aluminium accumulation in brains of renal dialysis patients. Lancet 1994; 343(8904): 993-7.
        [http://dx.doi.org/10.1016/S0140-6736(94)90124-4] [PMID: 7909090]

[66]   Sánchez-Iglesias S, Méndez-Alvarez E, Iglesias-González J, *et al.* Brain oxidative stress and selective behaviour of aluminium in specific areas of rat brain: potential effects in a 6-OHDA-induced model of Parkinson's disease. J Neurochem 2009; 109(3): 879-88.
        [http://dx.doi.org/10.1111/j.1471-4159.2009.06019.x] [PMID: 19425176]

[67]   Uversky VN, Li J, Bower K, Fink AL. Synergistic effects of pesticides and metals on the fibrillation of α-synuclein: implications for Parkinson's disease. Neurotoxicology 2002; 23(4-5): 527-36.
        [http://dx.doi.org/10.1016/S0161-813X(02)00067-0] [PMID: 12428725]

[68]   Shi-liang X, Miao L, Bing S, Chong-sheng B, Ji-hong Z, Yan-fei L. Effects of sub-chronic Aluminium exposure on renal pathologic structure in rats. J Northeast Agric Univ 2013; 20(1): 49-52. [English Edition].
        [http://dx.doi.org/10.1016/S1006-8104(13)60008-2]

[69]   Nehru B, Anand P. Oxidative damage following chronic aluminium exposure in adult and pup rat

brains. J Trace Elem Med Biol 2005; 19(2-3): 203-8.
[http://dx.doi.org/10.1016/j.jtemb.2005.09.004] [PMID: 16325537]

[70]   Lukiw WJ, Percy ME, Kruck TP. Nanomolar Aluminium induces pro-inflammatory and pro-apoptotic gene expression in human brain cells in primary culture. J Inorg Biochem 2005; 99(9): 1895-8.
[http://dx.doi.org/10.1016/j.jinorgbio.2005.04.021] [PMID: 15961160]

[71]   Verstraeten SV, Aimo L, Oteiza PI. Aluminium and lead: molecular mechanisms of brain toxicity. Arch Toxicol 2008; 82(11): 789-802.
[http://dx.doi.org/10.1007/s00204-008-0345-3] [PMID: 18668223]

[72]   Farhat SM, Mahboob A, Iqbal G, Ahmed T. Aluminium-induced cholinergic deficits in different brain parts and its implications on sociability and cognitive functions in mouse. Biol Trace Elem Res 2017; 177(1): 115-21.
[http://dx.doi.org/10.1007/s12011-016-0856-3] [PMID: 27709498]

[73]   Verstraeten SV, Oteiza PI. Aluminium and Phosphatidylinositol-Specific-Phospholipase C.Encyclopedia of Metalloproteins. New York, NY: Springer New York 2013; pp. 24-32.
[http://dx.doi.org/10.1007/978-1-4614-1533-6_433]

[74]   Johnson VJ, Kim S-H, Sharma RP. Aluminium-maltolate induces apoptosis and necrosis in neuro-2a cells: potential role for p53 signaling. Toxicol Sci 2005; 83(2): 329-39.
[http://dx.doi.org/10.1093/toxsci/kfi028] [PMID: 15537749]

[75]   Campbell A. The potential role of aluminium in Alzheimer's disease. Nephrol Dial Transplant 2002; 17 (Suppl. 2): 17-20.
[http://dx.doi.org/10.1093/ndt/17.suppl_2.17] [PMID: 11904353]

[76]   Johnson VJ, Sharma RP. Aluminium disrupts the pro-inflammatory cytokine/neurotrophin balance in primary brain rotation-mediated aggregate cultures: possible role in neurodegeneration. Neurotoxicology 2003; 24(2): 261-8.
[http://dx.doi.org/10.1016/S0161-813X(02)00194-8] [PMID: 12606298]

[77]   Iqbal G, Ahmed T. Co-exposure of metals and high fat diet causes aging like neuropathological changes in non-aged mice brain. Brain Res Bull 2019; 147: 148-58.
[http://dx.doi.org/10.1016/j.brainresbull.2019.02.013] [PMID: 30807793]

[78]   Andrade V, Aschner M, Dos Santos AM. Neurotoxicity of metal mixtures Neurotoxicity of metals. Springer 2017; pp. 227-65.
[http://dx.doi.org/10.1007/978-3-319-60189-2_12]

[79]   Chen P, Miah MR, Aschner M. Metals and Neurodegeneration. F1000 Res 2016; 5: 5.
[http://dx.doi.org/10.12688/f1000research.7431.1] [PMID: 27006759]

[80]   Jan AT, Azam M, Siddiqui K, Ali A, Choi I, Haq QM. Heavy metals and human health: mechanistic insight into toxicity and counter defense system of antioxidants. Int J Mol Sci 2015; 16(12): 29592-630.
[http://dx.doi.org/10.3390/ijms161226183] [PMID: 26690422]

[81]   Murthy R, Saxena D, Lal B. Effects of pre-and postnatal combined exposure to Pb and Mn on brain development in rats. Ind Health 1983; 21(4): 273-9.
[http://dx.doi.org/10.2486/indhealth.21.273] [PMID: 6197396]

[82]   Inan-Eroglu E, Ayaz A. Is Aluminium exposure a risk factor for neurological disorders? Journal of research in medical sciences: The official journal of Isfahan University of Medical Sciences 2018; 23.

[83]   Zheng W, Aschner M, Ghersi-Egea J-F. Brain barrier systems: a new frontier in metal neurotoxicological research. Toxicol Appl Pharmacol 2003; 192(1): 1-11.
[http://dx.doi.org/10.1016/S0041-008X(03)00251-5] [PMID: 14554098]

[84]   Karri V, Schuhmacher M, Kumar V. Heavy metals (Pb, Cd, As and MeHg) as risk factors for cognitive dysfunction: A general review of metal mixture mechanism in brain. Environ Toxicol Pharmacol 2016; 48: 203-13.

[http://dx.doi.org/10.1016/j.etap.2016.09.016] [PMID: 27816841]

[85] Zheng W. Neurotoxicology of the brain barrier system: new implications. J Toxicol Clin Toxicol 2001; 39(7): 711-9.
[http://dx.doi.org/10.1081/CLT-100108512] [PMID: 11778669]

[86] Méndez-Armenta M, Ríos C. Cadmium neurotoxicity. Environ Toxicol Pharmacol 2007; 23(3): 350-8.
[http://dx.doi.org/10.1016/j.etap.2006.11.009] [PMID: 21783780]

[87] McCall MA, Gregg RG, Behringer RR, *et al.* Targeted deletion in astrocyte intermediate filament (Gfap) alters neuronal physiology. Proc Natl Acad Sci USA 1996; 93(13): 6361-6.
[http://dx.doi.org/10.1073/pnas.93.13.6361] [PMID: 8692820]

[88] Rai A, Maurya SK, Khare P, Srivastava A, Bandyopadhyay S. Characterization of developmental neurotoxicity of As, Cd, and Pb mixture: synergistic action of metal mixture in glial and neuronal functions. Toxicol Sci 2010; 118(2): 586-601.
[http://dx.doi.org/10.1093/toxsci/kfq266] [PMID: 20829427]

[89] van Exel E, de Craen AJ, Gussekloo J, *et al.* Association between high-density lipoprotein and cognitive impairment in the oldest old. Ann Neurol 2002; 51(6): 716-21.
[http://dx.doi.org/10.1002/ana.10220] [PMID: 12112077]

[90] Farr SA, Yamada KA, Butterfield DA, *et al.* Obesity and hypertriglyceridemia produce cognitive impairment. Endocrinology 2008; 149(5): 2628-36.
[http://dx.doi.org/10.1210/en.2007-1722] [PMID: 18276751]

[91] Razay G, Vreugdenhil A, Wilcock G. The metabolic syndrome and Alzheimer disease. Arch Neurol 2007; 64(1): 93-6.
[http://dx.doi.org/10.1001/archneur.64.1.93] [PMID: 17210814]

[92] Chen X, Hui L, Geiger JD. Role of LDL cholesterol and endolysosomes in amyloidogenesis and Alzheimer's disease. J Neurol Neurophysiol 2014; 5(5): 236.
[http://dx.doi.org/10.4172/2155-9562.1000236] [PMID: 26413387]

[93] McGrowder D, Riley C, Morrison EYSA, Gordon L. The role of high-density lipoproteins in reducing the risk of vascular diseases, neurogenerative disorders, and cancer. Cholesterol 2010.

[94] Agarwal S, Zaman T, Tuzcu EM, Kapadia SR. Heavy metals and cardiovascular disease: results from the National Health and Nutrition Examination Survey (NHANES) 1999-2006. Angiology 2011; 62(5): 422-9.
[http://dx.doi.org/10.1177/0003319710395562] [PMID: 21421632]

[95] Sanders AP, Mazzella MJ, Malin AJ, *et al.* Combined exposure to lead, cadmium, mercury, and arsenic and kidney health in adolescents age 12-19 in NHANES 2009-2014. Environ Int 2019; 131: 104993.
[http://dx.doi.org/10.1016/j.envint.2019.104993] [PMID: 31326826]

[96] Buhari O, Dayyab FM, Igbinoba O, Atanda A, Medhane F, Faillace RT. The association between heavy metal and serum cholesterol levels in the US population: National Health and Nutrition Examination Survey 2009-2012. Hum Exp Toxicol 2020; 39(3): 355-64.
[http://dx.doi.org/10.1177/0960327119889654] [PMID: 31797685]

[97] Lesser G, Kandiah K, Libow LS, *et al.* Elevated serum total and LDL cholesterol in very old patients with Alzheimer's disease. Dement Geriatr Cogn Disord 2001; 12(2): 138-45.
[http://dx.doi.org/10.1159/000051248] [PMID: 11173887]

[98] Weisgraber KH, Mahley RW. Human apolipoprotein E: the Alzheimer's disease connection. FASEB J 1996; 10(13): 1485-94.
[http://dx.doi.org/10.1096/fasebj.10.13.8940294] [PMID: 8940294]

[99] Padilla MA, Elobeid M, Ruden DM, Allison DB. An examination of the association of selected toxic metals with total and central obesity indices: NHANES 99-02. Int J Environ Res Public Health 2010; 7(9): 3332-47.

[http://dx.doi.org/10.3390/ijerph7093332] [PMID: 20948927]

[100] Tsaih S-W, Korrick S, Schwartz J, *et al.* Lead, diabetes, hypertension, and renal function: the normative aging study. Environ Health Perspect 2004; 112(11): 1178-82.
[http://dx.doi.org/10.1289/ehp.7024] [PMID: 15289163]

[101] Kristal-Boneh E, Coller D, Froom P, Harari G, Ribak J. The association between occupational lead exposure and serum cholesterol and lipoprotein levels. Am J Public Health 1999; 89(7): 1083-7.
[http://dx.doi.org/10.2105/AJPH.89.7.1083] [PMID: 10394320]

[102] Wojtczak-Jaroszowa J, Kubow S. Carbon monoxide, carbon disulfide, lead and cadmium--four examples of occupational toxic agents linked to cardiovascular disease. Med Hypotheses 1989; 30(2): 141-50.
[http://dx.doi.org/10.1016/0306-9877(89)90101-1] [PMID: 2682148]

[103] Samarghandian S, Azimi-Nezhad M, Shabestari MM, Azad FJ, Farkhondeh T, Bafandeh F. Effect of chronic exposure to cadmium on serum lipid, lipoprotein and oxidative stress indices in male rats. Interdiscip Toxicol 2015; 8(3): 151-4.
[http://dx.doi.org/10.1515/intox-2015-0023] [PMID: 27486375]

[104] Burkhead JL, Lutsenko S. The role of copper as a modifier of lipid metabolism. IntechOpen; 2013.

[105] Arrifano GPF, de Oliveira MA, Souza-Monteiro JR, *et al.* Role for apolipoprotein E in neurodegeneration and mercury intoxication. Front Biosci (Elite Ed) 2018; 10: 229-41.
[PMID: 28930615]

[106] Kumar NT, Liestøl K, Løberg EM, Reims HM, Mæhlen J. Apolipoprotein E allelotype is associated with neuropathological findings in Alzheimer's disease. Virchows Arch 2015; 467(2): 225-35.
[http://dx.doi.org/10.1007/s00428-015-1772-1] [PMID: 25898889]

[107] Dubelaar EJ, Verwer RW, Hofman MA, Van Heerikhuize JJ, Ravid R, Swaab DE. ApoE epsilon4 genotype is accompanied by lower metabolic activity in nucleus basalis of Meynert neurons in Alzheimer patients and controls as indicated by the size of the Golgi apparatus. J Neuropathol Exp Neurol 2004; 63(2): 159-69.
[http://dx.doi.org/10.1093/jnen/63.2.159] [PMID: 14989602]

[108] Ghebremedhin E, Schultz C, Braak E, Braak H. High frequency of apolipoprotein E ε4 allele in young individuals with very mild Alzheimer's disease-related neurofibrillary changes. Exp Neurol 1998; 153(1): 152-5.
[http://dx.doi.org/10.1006/exnr.1998.6860] [PMID: 9743577]

[109] Wierenga CE, Clark LR, Dev SI, *et al.* Interaction of age and APOE genotype on cerebral blood flow at rest. J Alzheimers Dis 2013; 34(4): 921-35.
[http://dx.doi.org/10.3233/JAD-121897] [PMID: 23302659]

[110] Koizumi K, Hattori Y, Ahn SJ, Buendia I, Ciacciarelli A, Uekawa K, *et al.* Apoε4 disrupts neurovascular regulation and undermines white matter integrity and cognitive function. Nat Commun 2018; 9(1): 1-11.
[http://dx.doi.org/10.1038/s41467-018-06301-2] [PMID: 29317637]

[111] Molloy DW, Standish TI, Nieboer E, Turnbull JD, Smith SD, Dubois S. Effects of acute exposure to Aluminium on cognition in humans. J Toxicol Environ Health A 2007; 70(23): 2011-9.
[http://dx.doi.org/10.1080/15287390701551142] [PMID: 17966072]

[112] Zhang L, Wang H, Abel GM, Storm DR, Xia Z. The Effects of Gene-Environment Interactions Between Cadmium Exposure and Apolipoprotein E4 on Memory in a Mouse Model of Alzheimer's Disease. Toxicol Sci 2020; 173(1): 189-201.
[http://dx.doi.org/10.1093/toxsci/kfz218] [PMID: 31626305]

[113] Bell RD, Winkler EA, Singh I, *et al.* Apolipoprotein E controls cerebrovascular integrity *via* cyclophilin A. Nature 2012; 485(7399): 512-6.
[http://dx.doi.org/10.1038/nature11087] [PMID: 22622580]

[114] Halliday MR, Pomara N, Sagare AP, Mack WJ, Frangione B, Zlokovic BV. Relationship between cyclophilin a levels and matrix metalloproteinase 9 activity in cerebrospinal fluid of cognitively normal apolipoprotein e4 carriers and blood-brain barrier breakdown. JAMA Neurol 2013; 70(9): 1198-200.
[http://dx.doi.org/10.1001/jamaneurol.2013.3841] [PMID: 24030206]

[115] Stewart WF, Schwartz BS, Simon D, Kelsey K, Todd AC. ApoE genotype, past adult lead exposure, and neurobehavioral function. Environ Health Perspect 2002; 110(5): 501-5.
[http://dx.doi.org/10.1289/ehp.02110501] [PMID: 12003753]

[116] Zhao N, Attrebi ON, Ren Y, et al. APOE4 exacerbates α-synuclein pathology and related toxicity independent of amyloid. Sci Transl Med 2020; 12(529): eaay1809.
[http://dx.doi.org/10.1126/scitranslmed.aay1809] [PMID: 32024798]

[117] Wright RO, Hu H, Silverman EK, et al. Apolipoprotein E genotype predicts 24-month bayley scales infant development score. Pediatr Res 2003; 54(6): 819-25.
[http://dx.doi.org/10.1203/01.PDR.0000090927.53818.DE] [PMID: 12930912]

[118] Xu H, Finkelstein DI, Adlard PA. Interactions of metals and Apolipoprotein E in Alzheimer's disease. Front Aging Neurosci 2014; 6: 121.
[http://dx.doi.org/10.3389/fnagi.2014.00121] [PMID: 24971061]

[119] Lee J-Y, Cho E, Kim T-Y, et al. Apolipoprotein E ablation decreases synaptic vesicular zinc in the brain. Biometals 2010; 23(6): 1085-95.
[http://dx.doi.org/10.1007/s10534-010-9354-9] [PMID: 20556483]

[120] Izaks GJ, Gansevoort RT, van der Knaap AM, Navis G, Dullaart RP, Slaets JP. The association of APOE genotype with cognitive function in persons aged 35 years or older. PLoS One 2011; 6(11): e27415.
[http://dx.doi.org/10.1371/journal.pone.0027415] [PMID: 22110642]

[121] Butterfield DA, Mattson MP. Apolipoprotein E and oxidative stress in brain with relevance to Alzheimer's disease. Neurobiol Dis 2020; 138: 104795.
[http://dx.doi.org/10.1016/j.nbd.2020.104795] [PMID: 32036033]

[122] Engstrom AK, Snyder JM, Maeda N, Xia Z. Gene-environment interaction between lead and Apolipoprotein E4 causes cognitive behavior deficits in mice. Mol Neurodegener 2017; 12(1): 1-23.
[PMID: 28049533]

[123] Mielke MM, Hagen CE, Wennberg AMV, et al. Association of plasma total tau level with cognitive decline and risk of mild cognitive impairment or dementia in the mayo clinic study on aging. JAMA Neurol 2017; 74(9): 1073-80.
[http://dx.doi.org/10.1001/jamaneurol.2017.1359] [PMID: 28692710]

[124] Lu X. Occupational exposure to Aluminium and cognitive impairment Neurotoxicity of Aluminium. Springer 2018; pp. 85-97.

[125] Mohammed RS, Ibrahim W, Sabry D, El-Jaafary SI. Occupational metals exposure and cognitive performance among foundry workers using tau protein as a biomarker. Neurotoxicology 2020; 76: 10-6.
[http://dx.doi.org/10.1016/j.neuro.2019.09.017] [PMID: 31593711]

[126] Guilarte TR. Manganese neurotoxicity: new perspectives from behavioral, neuroimaging, and neuropathological studies in humans and non-human primates. Front Aging Neurosci 2013; 5: 23.
[http://dx.doi.org/10.3389/fnagi.2013.00023] [PMID: 23805100]

[127] Lu X, Liang R, Jia Z, et al. Cognitive disorders and tau-protein expression among retired Aluminium smelting workers. J Occup Environ Med 2014; 56(2): 155-60.
[http://dx.doi.org/10.1097/JOM.0000000000000100] [PMID: 24451610]

[128] Kim AC, Lim S, Kim YK. Metal ion effects on Aβ and tau aggregation. Int J Mol Sci 2018; 19(1): 128.

[http://dx.doi.org/10.3390/ijms19010128] [PMID: 29301328]

[129]   Alquezar C, Felix JB, McCandlish E, *et al.* Heavy metals contaminating the environment of a progressive supranuclear palsy cluster induce tau accumulation and cell death in cultured neurons. Sci Rep 2020; 10(1): 569.
[http://dx.doi.org/10.1038/s41598-019-56930-w] [PMID: 31953414]

[130]   Sastre M, Ritchie CW, Hajji N. Metal ions in Alzheimer's disease brain. JSM Alzheimers Dis Relat Dement 2015; 2: 1014.

[131]   McLachlan D. Aluminium and the risk for Alzheimer's disease. Environmetrics 1995; 6(3): 233-75.
[http://dx.doi.org/10.1002/env.3170060303]

[132]   Chen W-T, Liao Y-H, Yu H-M, Cheng IH, Chen Y-R. Distinct effects of Zn2+, Cu2+, Fe3+, and Al3+ on amyloid-β stability, oligomerization, and aggregation: amyloid-β destabilization promotes annular protofibril formation. J Biol Chem 2011; 286(11): 9646-56.
[http://dx.doi.org/10.1074/jbc.M110.177246] [PMID: 21216965]

[133]   Kawahara M, Kato-Negishi M, Tanaka KI. Amyloids: Regulators of Metal Homeostasis in the Synapse. Molecules 2020; 25(6): 1441.
[http://dx.doi.org/10.3390/molecules25061441] [PMID: 32210005]

CHAPTER 3

Role of aluminium in Post-Translational Modifications and Neurological Disorders

Saadia Zahid[1,*], **Sanila Amber**[1] and **Fatima Javed Mirza**[1]

[1] *Neurobiology Laboratory, Atta ur Rahman School of Applied Biosciences, National University of Sciences and Technology, Islamabad, Pakistan*

Abstract: Increased exposure or elevated levels of aluminium(Al) in humans cause various detrimental pathological processes especially affecting the central nervous system. Al-induced neurotoxicity predominantly leads to impaired motor coordination, cognition and learning and memory deficits. Significant association of chronic Al exposure with several neurological disorders, including Alzheimer's disease (AD), amyotrophic lateral sclerosis (ALS) Parkinson's disease (PD) and multiple sclerosis (MS) is evident where it instigates aberrant expression of various proteins *via* alterations in post-translational modifications (PTMs). In depth understanding of mechanism of action of Al, effect of altered PTMs and their detection methods is essential to revert anomalies induced by Al in these neurological disorders. The present chapter will attempt to summarize the role of Al in modulation of significant PTMs including phosphorylation, methylation, oxidation, ubiquitination and provide insights into its involvement in various neurological disorders.

Keywords: Aluminium, Alzheimer's Disease, Amyotrophic lateral Sclerosis, Multiple Sclerosis, Parkinson's Disease, Post-Translational Modifications.

INTRODUCTION

Aluminium (Al) is a highly abundant metal in the earth's crust and has been in use by mankind for centuries. It is widely utilized for food preservation, cans, kitchen utensils, and vaccine adjuvants, *etc*. Because of its widespread availability, both in environment and foodstuff, human exposure is almost unavoidable. Increased exposure or elevated levels of Al in humans may cause various detrimental pathological processes especially affecting the central nervous system (CNS). Elevated levels of Al in the brain are associated with mitochondrial   dysfunction,

---

* **Corresponding author Saadia Zahid**: Neurobiology Laboratory, Atta ur Rahman School of Applied Biosciences, National University of Sciences and Technology, Islamabad, Pakistan; Tel:+92 51 90856159;
E-mail: saadia.zahid@asab.nust.edu.pk

Touqeer Ahmed (Ed.)

apoptosis, lipid peroxidation, oxidative stress, protein misfolding, and neurotoxicity leading to neurodegeneration (Fig. **1**).

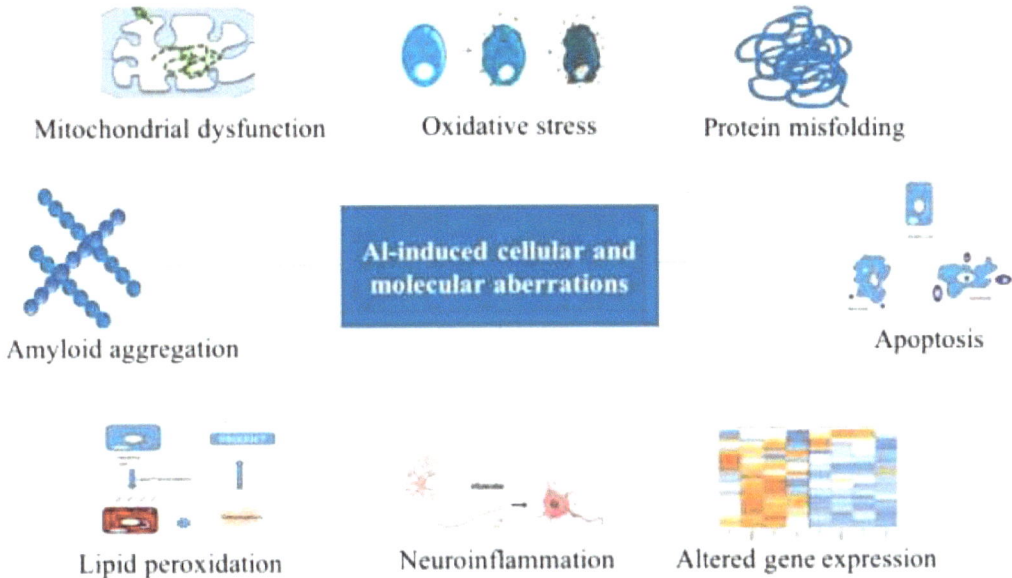

Fig. (1).   Diagrammatic representation of Aluminium-induced cellular and molecular aberrations.

1. ALUMINIUM MEDIATED OXIDATIVE STRESS

Although oxygen is essential for the survival of living organisms and serves to meet the energy needs of the biological tissues, it produces an increase in the levels of free radicals, resulting in toxic effects on the tissue. Hyperoxia and the deteriorative effects of reactive oxygen species (ROS) have also been attributed to neurotoxicity as brain tissues have a high consumption rate of oxygen and low activity of antioxidant enzymes. Partially reduced forms of oxygen or ROS exist in different varieties, including hydrogen peroxide (H_2O_2), superoxide (O_2^-) and the hydroxyl radical ($OH\bullet$) and react with the biological molecules altering their function and causing cell damage and death [1]. Oxidative stress occurs as a consequence of increased ROS brought about by an imbalance in the production of ROS and the levels of antioxidants, ultimately affecting the cells adversely. It has been implicated in various neurological disorders including Alzheimer's disease (AD), Huntington's disease (HD) and amyotrophic lateral sclerosis (ALS) [2].

Al, despite its low redox potential, causes oxidative damage through various mechanisms. It binds to the negatively charged phospholipids containing polyunsaturated fatty acids in the neuronal cell membranes rendering them more susceptible to the effect of ROS. It also causes the production of ROS and Fe^{3+} through the stimulation of iron-initiated lipid peroxidation in the Fenton reaction. Additionally, it raises the oxidative capacity of superoxide (O_2^-) by neutralizing it to form an $Al-O_2^-$ complex [3].

Numerous studies conducted on animal models have revealed that prolonged exposure to Al has resulted in oxidative damage that was more evident in the prefrontal cortex, cerebellum, hippocampus and brainstem. It was also found that, in addition to a significant increase in lipid peroxidation, Al also declines the cellular levels of antioxidant enzymes like superoxide dismutase, catalase and glutathione peroxidase and transferase. A sharp decrease in the mitochondrial activity is also associated with the detrimental oxidative damage induced by Al which causes disruption in the mitochondrial bioenergetics and reduces the respiratory efficiency and mitochondrial capacity [4].

2. ALUMINIUM INDUCED NEUROTOXICITY

Al exhibits neurotoxic effects which could be attributed to its high affinity to proteins and the ability to crosslink them [5]. An initial account of these neurotoxic effects was noticed in dialysis patients who were subjected to a dialysate enriched with Al salts. These patients not only exhibited increased concentrations of Al in the plasma and brain tissue but also showed symptoms of neurotoxicity like disorientation, memory loss which eventually led to dementia. Al also affects the hippocampal calcium signal pathways, thereby disrupting the neuronal plasticity and causing loss of memory and cognition [6].

3. ALUMINIUM AND NEUROLOGICAL DISORDERS

3.1. Aluminium and Alzheimer's Disease

Alzheimer's disease is a progressive neurodegenerative disorder characterized by the formation of amyloid beta plaques (Aβ) and neurofibrillary tangles (NFTs) and manifested by the loss of memory and cognition. The possibility of a role of Al in AD arose in 1965 with the accidental finding of neuronal changes involving neurofibrillary degeneration in the brain of rabbits injected with Al salts [7]. The cholinotoxic activity of Al coupled with the promotion of apoptotic neuronal loss makes it a highly potent neurotoxin inducing neurodegeneration and loss of learning and memory.

Al accumulates in the nerve cell nuclei in the AD brain predominantly in the hippocampus and presents in abundance in the senile plaques consisting of Aβ peptide aggregates [8]. It promotes the formation of Aβ plaques through an increase in the expression of the precursor (APP) of the Aβ protein as well as the levels of β-40 and β-42 fragments in the brain and enhances their aggregation. It also induces phosphorylation thereby promoting misfolding and aggregation of highly phosphorylated tau protein and alters synaptic plasticity [9].

3.2. Aluminium and other Neurological Disorders

Al is associated with several other neurological disorders including Parkinson's disease (PD), multiple sclerosis (MS) and amyotrophic lateral sclerosis (ALS). Occupational exposure with Al and other heavy metals like iron, lead, mercury, copper and manganese has been found to double the risk of PD. Elevated levels of Al substantially accelerates the fibrillization of α-synuclein, a major component of Lewy bodies and contributes to cerebellar dysfunction in PD. In addition, the hyperphosphorylation of tau protein, present in synaptic-enriched fractions in PD, is dependent on aggregated α-synuclein, indicating a synergistically modulated pathogenicity mechanism where Al drastically promotes the aggregation of hyperphosphorylated tau.

Similarly, the accumulation of Al along with other toxic metals such as cobalt, zinc manganese and copper is evident in CSF of ALS patients. Marked motor neuron degeneration with NFT formation was also observed in mice post Al exposure, that resembles ALS pathology [10].

Moreover, Al is associated with the neuropathology of MS including plaque like structures and corpora amylacea. In MS, the incidence of Al is equally high in different brain regions, both in intracellular and extracellular locations [11].

4. POST TRANSLATIONAL MODIFICATIONS

Proteins play a prominent role in cellular physiology therefore; proteome of an individual is rather complex and primarily expanded following translation. PTMs are common in eukaryotes where various enzymes are solely committed to performing modifications of the proteins. PTMs such as phosphorylation, acetylation, methylation, palmitoylation, glycosylation, ubiquitination, sulfation, nitrosylation, hydroxylation, oxidation, and glycation alter the structure and function of proteins thus changing the physiology of the cell [12 - 21]. Although PTMs such as siderophorylation, glypiation, AMPylation, neddylation and cholesteroylation are rare but have substantial impact on protein function and structure [22]. Some important eukaryotic PTMs, their target residues and enzymes used to catalyze these modifications are presented in Table **1** .

Table 1. Eukaryotic PTMs, their target residues and enzymes involved.

Sr. No.	PTMs	Mass Changes (Da)	Target Amino Acid(s)	Enzymes Involved	Examples	References
1.	Phosphorylation	80	Serine, tyrosine, threonine	Phosphatases, kinases	Protein kinase A	[12]
2.	Acetylation	42	Lysine	Acetyltransferases, deacetylases	Histone lysine N-acetylation	[12]
3.	Myristoylation	210	Glysine	N-Myristoyltransferases	Src kinase	[21]
4.	Methylation	14	Lyine, arganine	Methyltransferases, demethylases	Histone lysine monomethylation	[13]
5.	Palmitoylation	238	Cysteine	acyl-protein thioesterases, DHHC protein acyltransferases	Ras GTPase	[14]
6.	Sulfation	80	Tyrosine	Sulfatases, desulfatases	CCR5 receptor	[17]
7.	Nitrosylation	29	Cysteine, methionine	Nitrosylases, denitrosylases	Tubulin	[18]
8.	Hydroxylation	16	Glycine, proline, lysine	Hydroxylases	HIF1-α, Collagen	[19]
9.	Glycosylation	> 2000	Serine, threonine, Asparagine	Glycosyltransferases, deglycosylases	Mucin, Transferrin	[15]
10.	Oxidation	15.9	Cysteine, methionine	Oxidases, peroxidases, thioredoxin, glutathione	Thioredoxin peroxidase B	[20]
11.	Ubiquitination	114	Lysine	Ubiquitin activating, and conjugating enzymes, deubiquitinases, ubiquitin ligases	p53	[16]

5. ALUMINIUM AND POST TRANSLATIONAL MODIFICATIONS

Chronic Al exposure leads to aberrant expression of various functionally important proteins *via* alterations in their PTMs. Al significantly modulates several PTMs including phosphorylation, methylation, oxidation, ubiquitination *etc* (Fig. **2**, Table **2**).

Table 2. Different compounds of aluminium, their dosage, route of administration and effects on PTMs.

Sr. No.	Compound	Dosage and Route of Administration	Model	PTM and Target Protein	Reference
1.	AlCl$_3$	100 mg/kg, Oral gavage	Male albino Wistar rats	Tau hyper-phosphorylation, GSK-3β and Akt de-phosphorylation	[24]
2.	Al^{3+}	250 and 500 μM, Rat brain in-situ	Human autopsy samples and Wistar rats	Tau hyper-phosphorylation	[26]
3.	AlCl$_3$	0.2%, 0.4% and 0.6%, Oral	Pregnant Wistar rats and their neonates	CREB phosphorylation	[27]
4.	AlCl$_3$	1.6 mg/kg, Oral/drinking water	Male Wistar rats	Decrease in PP2A phosphatase activity, Tau hyper-phosphorylation	[28]
5.	AlCl$_3$	2, 12, and 72 mg/kg, Oral/drinking water	Male Sprague Dawley rats	Reduction in H3K4me3 expression	[32]
6.	Al		Aluminium factory workers	Decrease in H3K4me3 and increase in H3K9me2 and H3K27me3 methylation	[32]
7.	Al$_2$(SO$_4$)$_3$	1000 ppb, Oral/drinking water	Geriatric rats	Decrease in catalase and increase in SOD, GPx and GSH-r expression	[36]
8.	Al	2 mg/Kg diet	Tg2576	Increase in isoprostane levels and Aβ expression	[40]
9.	Al$_2$(C$_4$H$_4$O$_6$)$_3$	Subcutaneous	Rabbit brain	MAP2 ubiquitination	[44]
10.	Al$_2$ ClH$_7$O$_6$	10^{-4} M	SH-SY5Y cells	Phosphorylation and ubiquitination of ERα	[45]
11.	Aβ-Al complex	10 μM	SH-SY5Y cells	Reduced expression of UCHL3	[46]

5.1. Aluminium and Phosphorylation

Protein phosphorylation is the most common type of PTM that acts in a reversible manner. Approximately, one third of the proteins undergo phosphorylation in humans and intricate balance in maintained *via* action of kinases and phosphatases. Kinases mediate the phosphorylation by transferring phosphoryl group from adenosine triphosphate (ATP) to the threonine, serine or tyrosine residues of target protein. Phosphorylation changes the protein conformation which in turns modifies the catalytic activity of protein and causes it to aggregate or misfold, activate or inactivate the protein or recruit other binding partners. These changes lead to variations in protein function and signaling process(s).

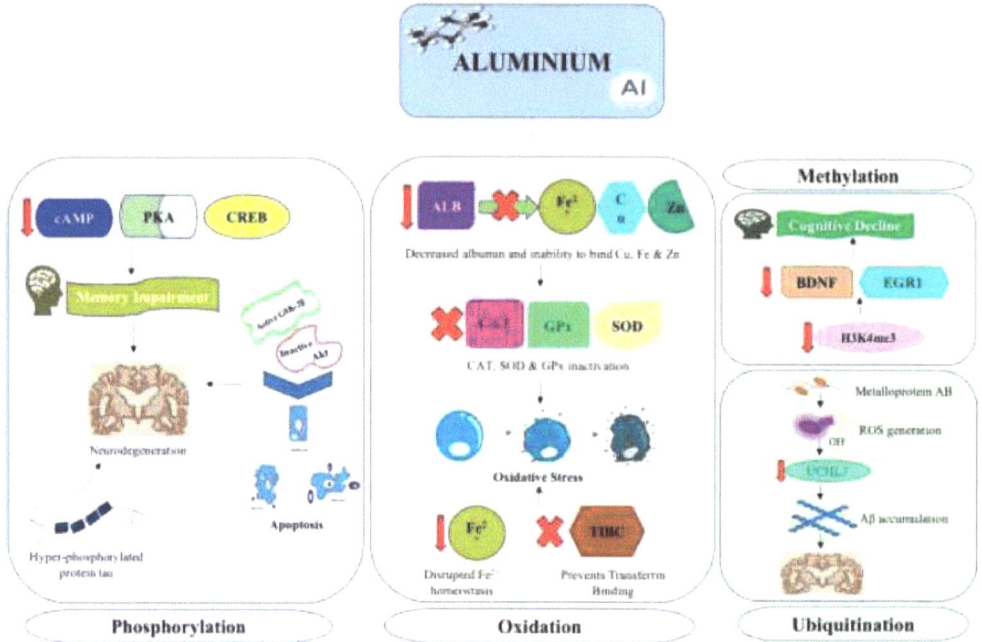

Fig. (2). Schematic representation of deleterious effects of aluminiumon phosphorylation, oxidation, methylation and ubiquitination.

Perturbance of protein phosphorylation is indicated in diabetes, cancers and neurological disorders. Etiology of the abnormal phosphorylation is complicated however, mutations and/or abnormal processing of kinases and phosphatases is an important contributing factor [23]. Therefore, the detailed knowledge of these enzyme repertoire is essential to present the broader picture of deregulated phosphorylation. This knowledge is essential to revert the aberrant phosphorylation and treat various disorders. Abnormal expression of different kinases such as cyclin dependent kinase 5 (CDK-5) and glycogen synthase kinase 3 beta (GSK-3β) could be an outcome of heavy metal toxicity. Heavy metals like Al elevates the expression of CDK-5 which in turn increases the expression of phosphorylated tau [24]. Microtubule associated protein tau (MAPT), a microtubule stabilizing protein maintains the structure of neurons. Abnormal phosphorylation detaches tau from microtubules and promotes the aggregation in the form of paired helical filaments (PHFs) that eventually lead to the accumulation of neurofibrillary tangles (NFTs). NFTs destabilize the synaptic connections between neurons and promote neuronal damage and memory impairment; important hallmarks of AD [25]. Thus, by regulating the expression of CDK-5, Al indirectly hyper phosphorylates tau and contributes the AD pathophysiology. Correspondingly, Al^{3+} ions intensively phosphorylated 67 kDa protein tau whereas inhibited the phosphorylation of 30 kDa, 55 kDa and 94 kDa isoforms of MAP in the rat brain [26].

Along with the hyper-phosphorylation of tau, Al causes reduced phosphorylation of cAMP response element-binding protein (CREB). Impairment in CREB phosphorylation disturbs cAMP-PKA-CREB pathway which then interferes with long term memory (LTM) formation [27]. The inactivation of protein kinase B (PKB or Akt) in the cortex and hippocampus of male Wistar rats following $AlCl_3$ administration is a consequence of $AlCl_3$ induced dephosphorylation. Akt is involved in cell survival and proliferation, where its diminished expression leads to apoptosis. Concomitantly, Al declined the phosphorylation of GSK-3β that led to its activation. GSK-3β negatively regulates cell proliferation and neurogenesis. Taken together, activation and inactivation of GSK-3β and Akt, respectively, leads to neuronal death [24] which is evident in several disorders such as PD, HD and AD.

Al mediates both enzymatic and non-enzymatic phosphorylation. During enzyme driven phosphorylation, kinases such as protein kinase C (PKC) and CDK-5 transfer phosphate group to the target proteins. Non-enzymatic phosphorylation is governed by the formation of covalent bond of target protein with alpha and gamma phosphates of adenosine triphosphate (ATP) and guanosine triphosphate (GTP). Another possible mechanism for the regulation of protein phosphorylation with Al is the inhibition of phosphatases such as reduction in the expression of protein phosphatase 2 (PP2A). It inhibits the PP2A activity at the concentration of 1.6 mg/kg that is close to the high end of Al consumed by humans in urban regions [28].

5.2. Aluminium and Methylation

Protein methylation is the most common type of alkylation that changes the hydrophobicity of the proteins. Methyltransferases catalyze the transfer of methyl group from S-adenosyl methionine (SAM) to the oxygen or nitrogen of the target amino acid residues on substrate proteins therefore referred to as O- and N-methylation, respectively. Although mostly studied on histone proteins, methylation of non-histone proteins is also crucially important and responsible for signal transduction mediated by bone morphogenetic protein (BMP), Janus kinase (JAK)–signal transducer and activator of transcription (STAT), mitogen activated protein kinase (MAPK), Hippo and WNT signaling pathways [13]. Histone methylation occurs at N-terminal and modifies transcriptional switch that further changes downstream signaling cascade(s) [29]. Initially, it was considered irreversible however, the discovery of demethylases marks the reversibility of methylation process.

Histone methylation commonly occurs at different lysine residues of histone 3 (H3). Among H3 histones, H3K4, H3K79 and H3K36 cause transcriptional activation while H3K9 and H3K27 lead to inhibition of transcriptional activity [30]. Methylation of histones is affected by chronic Al exposure and contributes to pathology of various neurological disorders. Histone lysine tri-methyl state (H3K4me3) a stable epigenetic modification, fine tunes the synaptic plasticity and long-term memory formation and consolidation however Al accumulation substantially reduces the H3K4me3 levels and impairs cognitive abilities [31].

Occupational Al exposure in thermoelectric and electrolyticAluminium workers revealed changes in cognitive functions along with the decrease in H3K4me3 and increase in H3K9me2 and H3K27me3 methylation. These changes were correlated to the expression of early growth response protein 1 (EGR1) and brain derived neurotrophic factor (BDNF) [32]. EGR1 is known to maintain the learning and memory functions [33] whereas BDNF plays significant role in neural development, synaptic activity, regeneration and repair [34]. Subsequently, by regulating the methylation state, Al reduces the activity of BDNF and EGR1 and thereby learning and memory related functions.

5.3. Aluminium and Protein Oxidation

Proteins interact with reactive oxygen species (ROS) and free radicals to undergo oxidative modifications. These modifications hydroxylate the side chains of aliphatic amino acid and aromatic groups, oxidize methionine and sulfhydryl groups, cleave polypeptide chain and form cross links [35]. Aberrantly oxidized proteins form aggregates and become resistant to proteolysis thus threatening the cellular viability.

Heavy metals such as Al produce free radicals thus causing an imbalance between oxidants and antioxidants that eventually trigger oxidative stress. It also causes disparity in the levels of antioxidant enzymes which is an important source of oxidative damage. Al decrease the expression of catalase and increase the levels of superoxide dismutase (SOD), glutathione peroxidase (GPx) and glutathione reductase (GSH-r) [36] in geriatric rats. Catalase breaks down hydrogen peroxide (H_2O_2) into water and oxygen and prevents the cell from damage. Low levels of catalase by Al or any other toxic insult increase the oxidative stress and escalate the likelihood of type 2 diabetes and obesity [37, 38]. Reduction in catalase activity may precede the increase in GPx and GSH-r which represent a compensatory mechanism to cope with oxidative stress [39]. Apart from this, Al administration increased isoprostane levels, Aβ expression and plaque deposition in the brain of Tg2576 mice. This effect of Al was pronounced in APP overexpressing (Tg2576) mice but not in wild type mice which suggests the involvement of Aβ toxicity in the progression of Al- induced oxidation [40].

Al causes oxidative damage *via* different mechanisms. Trace elements such as Cu, Fe and Zn act as co-factors for various antioxidant enzymes thus repairing the damage caused by oxidation and protecting the cell. Cu, Fe and Zn are transported to their target location by protein albumin, which is decreased by Al. Therefore, these elements are unable to activate the antioxidant enzymes (catalase, SOD and GPx) and causing oxidative stress which further leads to atypical oxidation of proteins and lipids. Moreover, Al hinders Fe homeostasis and displaces it from transferrin due to which Fe enters into the blood stream where it reacts with oxygen and produces superoxide anions and hydroxyl radicals [41].

5.4. Aluminium and Ubiquitination

The conjugation of ubiquitin to the target proteins is known as ubiquitination. Ubiquitin is a regulatory protein that is ubiquitously located in the eukaryotes and binds to the N-terminus lysine of the substrate protein during the process of ubiquitination. One or more ubiquitin molecules bind to the target protein, therefore referred to as mono and poly-ubiquitination. Mono- ubiquitination affects endocytosis, membrane trafficking and viral budding whereas poly-ubiquitination leads to the degradation of substrate protein *via* 26S proteasome pathway [42]. Ubiquitination affects various cellular processes including muscular and neural degeneration, development and differentiation, apoptosis, neural network formation, ribosome biogenesis and stress response. Differential levels of protein ubiquitination affect normal physiology of the cell by augmenting or reducing the degradation of various protein molecules [43].

Al co-localizes with ubiquitin and differentially affects the process of ubiquitination. Chronic Aluminium tartrate exposure increased anti-ubiquitin staining in the neurons of lower brain stem nuclei along with neurofibrillary changes. These changes may indicate the role of Al in NFTs formation which eventually lead to neuronal death [44].

Al also exerts neurodegenerative effects *via* estrogen dependent pathway as indicated by decrease in estrogen receptor beta (ERβ) protein levels and increase in its mitochondrial accumulation following aluminium chlorohydrate administration in SH-SY5Y cells. ERβ regulates mitochondrial biogenesis, anti-apoptotic and antioxidant activity. Therefore,aluminium chlorohydrate mediated decline in ERβ, leads to cell death. Decreased ERβ levels may in part be regulated by aluminium chlorohydrate-dependent increase in estrogen receptor alpha (ERα).aluminium chlorohydrate down-regulates phosphorylated state of ERα which inhibits the ubiquitination assisted proteolytic degradation and causes its accumulation. Consequently, Al impedes estrogen dependent protection mechanism which further leads to breast cancer and neuronal disorders [45].

Binding of Al to metalloprotein Aβ, aggravated ROS production and reduced mRNA expression of ubiquitin thiolesterase (UCHL3). UCHL3 is the key element of ubiquitin proteasome system that hydrolyzes C terminal amides and esters of ubiquitin. It replenishes mono-ubiquitin pool and regulates protein degradation. It is also involved in working memory and spatial learning and UCHL3 deficiency leads to memory impairment. Also, reduction in UCHL3 activity diminishes Aβ clearance which leads to aggregation of this peptide; as indicated in AD. It is therefore speculated that reduction in UCHL3 *via* interaction of Al and Aβ contributes to AD pathology [46].

CONCLUDING REMARKS

Aluminium-induced alterations in different PTMs is of critical importance. In depth understanding of the effect of these PTMs, mechanism of action and detection methods is essential to revert aberrations in several neurological disorders and pathological conditions. The identification and characterization of these modifications will serve as a valuable tool to alleviate PTMs associated brain damage. Advancement in proteomic strategies will provide further evidence of substantial involvement of Al-induced PTM aberrations, helpful for the development of improved diagnostic and therapeutic approaches for neurodegenerative disorders.

CONSENT FOR PUBLICATION

Not Applicable.

CONFLICT OF INTEREST

The author declares no conflict of interest, financial or otherwise.

ACKNOWLEDGEMENTS

Declared none.

LIST OF ABBREVIATIONS

Aβ   Amyloid beta

AD	Alzheimer's disease
Al	Aluminium
ALS	Amyotrophic lateral sclerosis
APP	Amyloid precursor protein
BDNF	Brain derived neurotrophic factor
BMP	Bone morphogenetic proteins
CAT	Catalase
CDK-5	Cyclin dependent kinase 5
CREB	cAMP response element-binding protein
CSF	Cerebrospinal fluid
EGR1	early growth response protein 1
ERα	Estrogen receptor alpha
ERβ	Estrogen receptor beta
GPx	Glutathione peroxidase
HD	Huntington's disease
JAK-STAT	Janus Kinase/Signal Transducer and Activator of Transcription
MAPK	Mitogen-activated protein kinase
MAPT	Microtubule associated protein tau
MS	Multiple sclerosis
NFTs	Neurofibrillary tangles
PHFs	Paired helical filaments
PTMs	Post translational modifications
ROS	Reactive oxygen species
SOD	Superoxide dismutase
UCHL3	Ubiquitin thiolesterase

REFERENCES

[1] Halliwell B. Oxidative stress and neurodegeneration: where are we now? J Neurochem 2006; 97(6): 1634-58.
[http://dx.doi.org/10.1111/j.1471-4159.2006.03907.x] [PMID: 16805774]

[2] Love S, Jenner P. Oxidative stress in brain ischemia. Brain Pathol 1999; 9(1): 119-31.
[http://dx.doi.org/10.1111/j.1750-3639.1999.tb00214.x] [PMID: 9989455]

[3] Exley C. The pro-oxidant activity of aluminium. Free Radic Biol Med 2004; 36(3): 380-7.
[http://dx.doi.org/10.1016/j.freeradbiomed.2003.11.017] [PMID: 15036357]

[4] Iglesias-González J, Sánchez-Iglesias S, Beiras-Iglesias A, Méndez-Álvarez E, Soto-Otero R. Effects of aluminium on rat brain mitochondria bioenergetics: an *in vitro* and in vivo study. Mol Neurobiol 2017; 54(1): 563-70.
[http://dx.doi.org/10.1007/s12035-015-9650-z] [PMID: 26742531]

[5] Klotz K, Weistenhöfer W, Neff F, Hartwig A, van Thriel C, Drexler H. The health effects of aluminium exposure. Dtsch Arztebl Int 2017; 114(39): 653-9.
[PMID: 29034866]

[6] Nday CM, Drever BD, Salifoglou T, Platt B. Aluminium interferes with hippocampal calcium signaling in a species-specific manner. J Inorg Biochem 2010; 104(9): 919-27.
[http://dx.doi.org/10.1016/j.jinorgbio.2010.04.010] [PMID: 20510457]

[7] Klatzo I, Wisniewski H, Streicher E. Experimental production of neurofibrillary degeneration: I. Light microscopic observations. J Neuropathol Exp Neurol 1965; 24(2): 187-99.
[http://dx.doi.org/10.1097/00005072-196504000-00002] [PMID: 14280496]

[8] Yumoto S, Kakimi S, Ishikawa A. Colocalization of aluminiumand iron in nuclei of nerve cells in brains of patients with Alzheimer's disease. J Alzheimers Dis 2018; 65(4): 1267-81.
[http://dx.doi.org/10.3233/JAD-171108] [PMID: 30149443]

[9] Bolognin S, Messori L, Drago D, Gabbiani C, Cendron L, Zatta P. Aluminium, copper, iron and zinc differentially alter amyloid-Aβ(1-42) aggregation and toxicity. Int J Biochem Cell Biol 2011; 43(6): 877-85.
[http://dx.doi.org/10.1016/j.biocel.2011.02.009] [PMID: 21376832]

[10] Tanridag T, Coskun T, Hürdag C, Arbak S, Aktan S, Yegen B. Motor neuron degeneration due to aluminium deposition in the spinal cord: a light microscopical study. Acta Histochem 1999; 101(2): 193-201.
[http://dx.doi.org/10.1016/S0065-1281(99)80018-X] [PMID: 10335362]

[11] Mold M, Chmielecka A, Rodriguez MRR, et al. Aluminium in brain tissue in multiple sclerosis. Int J Environ Res Public Health 2018; 15(8): 1777.
[http://dx.doi.org/10.3390/ijerph15081777] [PMID: 30126209]

[12] Marcelli S, Corbo M, Iannuzzi F, et al. The Involvement of post-translational modifications in Alzheimer's disease. Curr Alzheimer Res 2018; 15(4): 313-35.
[http://dx.doi.org/10.2174/1567205014666170505095109] [PMID: 28474569]

[13] Biggar KK, Li SS. Non-histone protein methylation as a regulator of cellular signalling and function. Nat Rev Mol Cell Biol 2015; 16(1): 5-17.
[http://dx.doi.org/10.1038/nrm3915] [PMID: 25491103]

[14] Guan X, Fierke CA. Understanding protein palmitoylation: biological significance and enzymology. Sci China Chem 2011; 54(12): 1888-97.
[http://dx.doi.org/10.1007/s11426-011-4428-2] [PMID: 25419213]

[15] Maverakis E, Kim K, Shimoda M, et al. Glycans in the immune system and The Altered Glycan Theory of Autoimmunity: a critical review. J Autoimmun 2015; 57(57): 1-13.
[http://dx.doi.org/10.1016/j.jaut.2014.12.002] [PMID: 25578468]

[16] Peng J, Schwartz D, Elias JE, et al. A proteomics approach to understanding protein ubiquitination. Nat Biotechnol 2003; 21(8): 921-6.
[http://dx.doi.org/10.1038/nbt849] [PMID: 12872131]

[17] Kehoe JW, Bertozzi CR. Tyrosine sulfation: a modulator of extracellular protein-protein interactions. Chem Biol 2000; 7(3): R57-61.
[http://dx.doi.org/10.1016/S1074-5521(00)00093-4] [PMID: 10712936]

[18] Nakamura T, Lipton SA. Protein S-nitrosylation as a therapeutic target for neurodegenerative diseases. Trends Pharmacol Sci 2016; 37(1): 73-84.
[http://dx.doi.org/10.1016/j.tips.2015.10.002] [PMID: 26707925]

[19] Lothrop AP, Torres MP, Fuchs SM. Deciphering post-translational modification codes. FEBS Lett 2013; 587(8): 1247-57.
[http://dx.doi.org/10.1016/j.febslet.2013.01.047] [PMID: 23402885]

[20]   Chung HS, Wang SB, Venkatraman V, Murray CI, Van Eyk JE. Cysteine oxidative posttranslational modifications: emerging regulation in the cardiovascular system. Circ Res 2013; 112(2): 382-92.
[http://dx.doi.org/10.1161/CIRCRESAHA.112.268680] [PMID: 23329793]

[21]   Omenn GS, Lane L, Lundberg EK, Beavis RC, Overall CM, Deutsch EW. Metrics for the Human Proteome Project 2016: progress on identifying and characterizing the human proteome, including post-translational modifications. J Proteome Res 2016; 15(11): 3951-60.
[http://dx.doi.org/10.1021/acs.jproteome.6b00511] [PMID: 27487407]

[22]   Basak S, Lu C, Basak A. Post-translational protein modifications of rare and unconventional types: implications in functions and diseases. Curr Med Chem 2016; 23(7): 714-45.
[http://dx.doi.org/10.2174/0929867323666160118095620] [PMID: 26778322]

[23]   Via A, Zanzoni A. A prismatic view of protein phosphorylation in health and disease. Front Genet 2015; 6: 131.
[http://dx.doi.org/10.3389/fgene.2015.00131] [PMID: 25904935]

[24]   Ahmad Rather M, Justin-Thenmozhi A, Manivasagam T, Saravanababu C, Guillemin GJ, Essa MM. Asiatic acid attenuated aluminium chloride-induced tau pathology, oxidative stress and apoptosis *via* AKT/GSK-3β signaling pathway in wistar rats. Neurotox Res 2019; 35(4): 955-68.
[http://dx.doi.org/10.1007/s12640-019-9999-2] [PMID: 30671870]

[25]   Iqbal K, Liu F, Gong CX. Tau and neurodegenerative disease: the story so far. Nat Rev Neurol 2016; 12(1): 15-27.
[http://dx.doi.org/10.1038/nrneurol.2015.225] [PMID: 26635213]

[26]   Shevtsov PN, Shevtsova EF, Savushkina OK, Burbaeva GS, Bachurin SO. Influence of Al^{3+}, Fe^{3+} and Zn^{2+} ions on phosphorylation of tubulin and microtubule-associated proteins of rat brain. Bull Exp Biol Med 2018; 165(4): 512-5.
[http://dx.doi.org/10.1007/s10517-018-4206-7] [PMID: 30121922]

[27]   Zhang L, Jin C, Lu X, *et al.* Aluminium chloride impairs long-term memory and downregulates cAMP-PKA-CREB signalling in rats. Toxicology 2014; 323: 95-108.
[http://dx.doi.org/10.1016/j.tox.2014.06.011] [PMID: 24973631]

[28]   Walton JR. An aluminium-based rat model for Alzheimer's disease exhibits oxidative damage, inhibition of PP2A activity, hyperphosphorylated tau, and granulovacuolar degeneration. J Inorg Biochem 2007; 101(9): 1275-84.
[http://dx.doi.org/10.1016/j.jinorgbio.2007.06.001] [PMID: 17662457]

[29]   Kouzarides T. Chromatin modifications and their function. Cell 2007; 128(4): 693-705.
[http://dx.doi.org/10.1016/j.cell.2007.02.005] [PMID: 17320507]

[30]   Upadhyay AK, Cheng X. Dynamics of histone lysine methylation: structures of methyl writers and erasers. Prog Drug Res 2011; 67: 107-24.
[http://dx.doi.org/10.1007/978-3-7643-8989-5_6] [PMID: 21141727]

[31]   Wang F, Kang P, Li Z, Niu Q. Role of MLL in the modification of H3K4me3 in aluminium-induced cognitive dysfunction. Chemosphere 2019; 232: 121-9.
[http://dx.doi.org/10.1016/j.chemosphere.2019.05.099] [PMID: 31152896]

[32]   Pan B, Zhou Y, Li H, *et al.* Relationship between occupational aluminium exposure and histone lysine modification through methylation. J Trace Elem Med Biol 2020; 61: 126551.
[http://dx.doi.org/10.1016/j.jtemb.2020.126551] [PMID: 32470791]

[33]   Cheval H, Chagneau C, Levasseur G, *et al.* Distinctive features of Egr transcription factor regulation and DNA binding activity in CA1 of the hippocampus in synaptic plasticity and consolidation and reconsolidation of fear memory. Hippocampus 2012; 22(3): 631-42.
[http://dx.doi.org/10.1002/hipo.20926] [PMID: 21425206]

[34]   Campos C, Rocha NBF, Lattari E, Nardi AE, Machado S. Exercise induced neuroplasticity to enhance therapeutic outcomes of cognitive remediation in schizophrenia: analyzing the role of brain derived

neurotrophic factor. CNS Neurol Disord Drug Targets 2017; 16(6): 638-51.
[http://dx.doi.org/10.2174/1871527315666161223142918] [PMID: 28017130]

[35]   Santos AL, Lindner AB. Protein posttranslational modifications: roles in aging and age-related disease. Oxid Med Cell Longev 2017; 2017: 5716409.
[http://dx.doi.org/10.1155/2017/5716409] [PMID: 28894508]

[36]   Muselin F, Gârban Z, Cristina RT, *et al.* Homeostatic changes of some trace elements in geriatric rats in the condition of oxidative stress induced by aluminium and the beneficial role of resveratrol. J Trace Elem Med Biol 2019; 55: 136-42.
[http://dx.doi.org/10.1016/j.jtemb.2019.06.013] [PMID: 31345351]

[37]   Góth L, Nagy T. Acatalasemia and diabetes mellitus. Arch Biochem Biophys 2012; 525(2): 195-200.
[http://dx.doi.org/10.1016/j.abb.2012.02.005] [PMID: 22365890]

[38]   Heit C, Marshall S, Singh S, *et al.* Catalase deletion promotes prediabetic phenotype in mice. Free Radic Biol Med 2017; 103: 48-56.
[http://dx.doi.org/10.1016/j.freeradbiomed.2016.12.011] [PMID: 27939935]

[39]   Shazia Q, Mohammad ZH, Rahman T, Shekhar HU. Correlation of oxidative stress with serum trace element levels and antioxidant enzyme status in Beta thalassemia major patients: a review of the literature. Anemia 2012; 2012: 270923.
[http://dx.doi.org/10.1155/2012/270923] [PMID: 22645668]

[40]   Praticò D, Uryu K, Sung S, Tang S, Trojanowski JQ, Lee VM. Aluminium modulates brain amyloidosis through oxidative stress in APP transgenic mice. FASEB J 2002; 16(9): 1138-40.
[http://dx.doi.org/10.1096/fj.02-0012fje] [PMID: 12039845]

[41]   Sahin Z, Ozkaya A, Yilmaz O, Yuce A, Gunes M. Investigation of the role of α-lipoic acid on fatty acids profile, some minerals (zinc, copper, iron) and antioxidant activity against aluminium-induced oxidative stress in the liver of male rats. Basic Clin Physiol Pharmacol 2017; 28(4): 355-61.
[http://dx.doi.org/10.1515/jbcpp-2015-0160] [PMID: 28306527]

[42]   Komander D. The emerging complexity of protein ubiquitination. Biochem Soc Trans 2009; 37(Pt 5): 937-53.
[http://dx.doi.org/10.1042/BST0370937] [PMID: 19754430]

[43]   Glickman MH, Ciechanover A. The ubiquitin-proteasome proteolytic pathway: destruction for the sake of construction. Physiol Rev 2002; 82(2): 373-428.
[http://dx.doi.org/10.1152/physrev.00027.2001] [PMID: 11917093]

[44]   Takeda M, Tatebayashi Y, Tanimukai S, Nakamura Y, Tanaka T, Nishimura T. Immunohistochemical study of microtubule-associated protein 2 and ubiquitin in chronically Aluminium-intoxicated rabbit brain. Acta Neuropathol 1991; 82(5): 346-52.
[http://dx.doi.org/10.1007/BF00296545] [PMID: 1722608]

[45]   Tsialtas I, Gorgogietas VA, Michalopoulou M, *et al.* Neurotoxic effects of aluminium are associated with its interference with estrogen receptors signaling. Neurotoxicology 2020; 77: 114-26.
[http://dx.doi.org/10.1016/j.neuro.2020.01.004] [PMID: 31945389]

[46]   Bolognin S, Zatta P, Lorenzetto E, Valenti MT, Buffelli M. β-Amyloid-aluminium complex alters cytoskeletal stability and increases ROS production in cortical neurons. Neurochem Int 2013; 62(5): 566-74.
[http://dx.doi.org/10.1016/j.neuint.2013.02.008] [PMID: 23416043]

CHAPTER 4

Effect of Aluminium on Synaptic Plasticity

Syeda Mehpara Farhat[1,*]

[1] *Department of Biological Sciences, National University of Medical Sciences, Rawalpindi-46000, Pakistan*

Abstract: Aluminium (Al) is the third most abundant metal in the earth's crust and it has long been associated with the pathogenesis of many neurological disorders. Recently, vast use of this metal in various industries and its elevated leaching from earth reservoirs, due to acid rain, has greatly increased human exposure to this metal. Due to the controversial nature of Al effects on the nervous system, it is important to thoroughly understand the effects of Al on neurological functions. This chapter is focused on understanding the effects of Al on the electrophysiological properties of neurons. The emphasis is on the effects of Al on synaptic plasticity, which is an important underlying mechanism in learning and memory, and voltage-gated ion channels. The evidence indicates that Al affects Long term potentiation (LTP), the most widely studied form of synaptic plasticity, *via* its effects on various signaling pathways.

Keywords: Aluminium, AMPA Receptor, Electrophysiological Variation, LTP, LTD, Metabotropic Glutamate Receptors, NMDA Receptor, Voltage-Gated Channel, Voltage dependent Calcium Channel, Neurotoxicity.

INTRODUCTION

Aluminium (Al) is the third most abundant element in the earth's crust and is a known neurotoxicant [1]. Al has been suggested to be neurotoxic based on different evidences from animal models, human studies and cell culture studies [2, 3]. Al gets easy access to the human body due to its addition to water purification systems, vaccine adjuvants and cosmetics [4, 5]. Al is also used as an additive in processed food [6]. Once in the body, Al is reported to cause damage to the blood brain barrier [7] and gets access to the brain *via* transferrin receptors and accumulates in the cortex and hippocampus [8]. The neurons are predisposed to toxic effects of Al accumulation due to their long life span [9]. The neurotoxic potential of Al has been reported in several studies and has been linked to various

---

[*] **Corresponding author Syeda Mehpara Farhat**: Department of Biological Sciences, National University of Medical Sciences, Rawalpindi-46000, Pakistan; Tel:??????; E-mail: mahpara.farhat@numspak.edu.pk

Touqeer Ahmed (Ed.)

neurodegenerative disorders, including multiple sclerosis [10], dialysis dementia [11], Parkinson's disease and Alzheimer's disease [12]. Although the causative role of Al in neurodegenerative disorders is controversial but common consensus is that Al may exacerbate the underlying events associated with neurodegenerative disorders [13].

Various studies on animal models have shown that, through its neurotoxic mechanisms, Al causes learning and memory deficits [14, 15]. The learning-related cellular changes might be either due to modifications at synapses or changes in the intrinsic properties of neurons [16] that may cause a decline in synaptic plasticity [17]. Synaptic plasticity refers to the changes in the strength of synaptic responses according to the neuronal activity [18] and it is the underlying process in memory formation *via* conversion of transient experiences to persistent memory [19].

Due to its positive oxidation state Al^{+3} has great affinity towards oxygen donors having a negative charge [16]. Therefore, the binding of Al to receptors and enzymes involved in neurotransmitter synthesis might affect neurotransmitter systems [20 - 22] and consequently result in synaptic plasticity impairment. These properties of Al cause neurotoxicity, leading to neurodegeneration and consequently learning and memory deficits [19]. The ability of Al to cause learning and memory deficits *via* its effects on synaptic plasticity has been known for more than two decades [23, 24] and is discussed in this chapter.

Effects of Al Accumulation on Synaptic Plasticity

With an increase or decrease in synaptic activity, the connections between neurons may strengthen or weaken. This change in the strength of the connection between neurons is referred to as synaptic plasticity [18]. The enhancement in the strength of synapses, in response to neuronal stimulation is referred to as long term potentiation (LTP), whereas the reduction in synaptic strength is termed as long term depression (LTD). Synaptic plasticity is a basic neural mechanism of learning and memory, which are the main functions that deteriorate due to Al accumulation in the brain. Al exposure is reported to affect both the early phase LTP, which is protein synthesis independent [25], and late phase LTP, which requires new protein synthesis, in the CA1 region of the hippocampus [26]. Al administration affects LTP, both *in vivo* and *in vitro*, in a concentration dependent manner [23]. Similarly, from the recordings performed on the dentate gyrus of hippocampus, it was observed that Al administration affects LTD along with LTP in neonatal rats [27] and in adult rats [28]. Al administration is also known to inhibit induction of tetra-ethyl ammonium (TEA)-induced synaptic enhancement of LTP, in a concentration and time dependent manner [24]. TEA is a potassium

(K^{+2}) channel blocker and is frequently used to study the mechanism of LTP, induced by electrical stimulation of afferent fibers [24]. These inhibitory effects of Al might be due to inhibition of Ca^{+2} conduction, which might interfere with Ca^{+2} dependent processes [24]. Moreover, the Al-induced impairment in synaptic plasticity might be due to increased apoptosis of hippocampal neurons [25].

Effect of Al on Ca^{+2} Channel and Ca^{+2} Signaling

During the development of synaptic plasticity, strong depolarization currents result in a higher influx of Ca^{+2} ions in the postsynaptic neuron. This Ca^{+2} ion completely displaces the Mg^{+2} ion, which blocks N-methyl-D-aspartate receptor (NMDAR), and this results in LTP induction *via* NMDAR activation. Weak depolarization of postsynaptic neuron leads to entry of a smaller amount of Ca^{+2} inside the cell and, therefore, partial replacement of Mg^{+2} ion, which leads to LTD induction. Thus the entry of Ca^{+2} into the cell is crucial for the induction of both LTP and LTD [29]. The voltage-dependent calcium channels (VDCC) are of prime importance because Ca^{+2} influx through these channels leads to activation of various events that result in the release of neurotransmitter glutamate from the presynaptic cells. This neurotransmitter consequently activates NMDAR and metabotropic glutamate receptor (mGluR) on postsynaptic cells for induction of LTP or LTD [16]. The Al inhibits Ca^{+2} influx through these channels [30], which leads to a reduction in the voltage-activated calcium channel current [31]. Moreover, due to the important role of mitochondria in the regulation of synaptic plasticity [32], irreversible inhibition of mitochondrial voltage-gated channel (VDAC) permeability by the micromolar quantity of Al [33], may also contribute to Al-induced synaptic plasticity inhibition. Furthermore, Al is also reported to affect high voltage activated (I_{HVA}) Ca^{+2} channels [34]. This effect of Al on Ca^{+2} channels might be due to the interaction of Al with binding sites within and outside these channels [31]. But these effects of Al on Ca^{+2} channels are not ubiquitous but different Al compounds act differently on these channels [35]. Moreover, the blockade of Ca^{+2} channels with Al is pH dependent and the extent of blockade increased with reduction in pH [36]. The application of Al is also reported to shift the current-voltage relationship towards depolarized voltage in cultured rat dorsal root ganglion neurons. But, this shift is Al concentration dependent and the magnitude varies in different cells [36]. The impairment in synaptic transmission has been observed due to inhibitory action of Al on voltage-gated Ca^{+2} channels [37]. Furthermore, Al acts as an antagonist of the enzymes containing Ca^{+2} and Mg^{+2} as these ions are replaced by Al and, therefore, the activity of Ca^{+2} dependent protein kinases is inhibited [38]. Thus the deficits in Ca^{+2} signaling might be the reason for Al induced learning and memory deterioration [39].

Effect of Aluminium on Receptors Involved in Synaptic Plasticity

There are multiple forms of synaptic plasticity, depending on the brain region and neuron type involved [16]. Extensive experimental investigations have been done on NMDAR-dependent LTP and LTD in the CA1 region of the hippocampus [16, 40]. In the hippocampus, the glutamatergic receptors, including NMDAR, metabotropic glutamate receptors (mGluR) and α-amino-3-hydroxy-5-me-hyl-4-isoxazole propionic acid receptor (AMPAR) are the major postsynaptic receptors involved in the emergence of synaptic plasticity [19]. The NMDAR-dependent synaptic plasticity is driven by Ca^{+2} influx, which occurs as a result of the activation of NMDAR [41]. The elevated Ca^{+2} level in the cell leads to activation of various synaptic proteins, including adenylate cyclase, protein kinase A, protein kinase C, calcineurin, Ca^{+2}/CAM dependent protein kinase II and protein phosphatase I. These proteins, in turn, alter AMPAR-related proteins. The AMPAR causes a further influx of Ca^{+2} ions and as a result, induction of synaptic plasticity takes place [42].

Administration of Al results in a reduction of the glutamate mediated currents involving NMDAR, AMPAR and other glutamate receptors [43]. Al exposure, in addition to reducing Ca^{+2} concentration, is also known to inhibit NMDA receptor α gene in rats during development [39]. Moreover, exposure of cultured hippocampal neurons with 37 and 74 µM Al for a period of 14 days was also observed to hamper the expression of NMDAR 1 in the cortex [44] and NMDAR 1A and NMDAR 2A/B in the hippocampus of rats [45]. In diabetic rats, it was found that Al administration causes a reduction in the expression of NMDAR 1 but upregulates expression of NMDAR 2B and, therefore, aggravates NMDAR signaling [46]. Moreover, rats born from females who were exposed to Al during pregnancy and lactation had reduced NMDAR 1 and NMDAR 2B expression and had impaired learning and memory [47]. Thus, Al has the ability to influence the NMDAR expression and its function that consequently causes disturbance in Ca^{+2} level of the cell, which leads to impairment in learning and memory [48]. The reduced expression and phosphorylation of NMDAR due to Al-induced upregulation of mGluR1 receptors is another mechanism through which Al hampers the maintenance and induction of LTP [49].

In addition to its effects on NMDAR, Al administration is also reported to reduce expression of AMPA receptor subunits, GluR1 and GluR2, in rat hippocampus in a dose-dependent manner [50]. Even at low concentration, Al is reported to manifest its effects at genetic level and the expression of AMPA receptor GluR1 and GluR2 genes were downregulated. However, expression of GluR3 gene is altered only on administration of high Al concentrations [51].

The role of inhibitory GABAergic receptors in synaptic plasticity has recently been reported and it is now known that the inhibitory synapses also undergo various forms of synaptic plasticity and add versatility to brain states [52]. The $GABA_A$ receptors are the main fast inhibitory neurotransmitters [53] and Al-induced alteration in the $GABA_A$ receptor function might be a contributing factor to cognitive deficits [54]. This reveals that Al may cause the disruption of synaptic plasticity *via* its effects on all the receptors involved in the induction and maintenance of synaptic plasticity.

Effect of Aluminium on Signaling Pathways Involved in Synaptic Plasticity

The effects of Al on cell signaling pathways are of particular importance (Fig. **1**). Al is reported to affect various signaling pathways, including phosphotidylinositol-specific phospholipase C (PI-PLC), G-protein dependent signaling pathways and protein kinase C (PKC) [55]. Wang *et al.* have reported that Al administration in rats reduces the activity of mitogen-activated protein kinase (MAPK), which is crucial for long-term memory formation [56]. Moreover, Al reduces the expression of extracellular signal-regulated kinases (ERK1/2) and Ca^{+2} calmodulin dependent protein kinase II (CaMKII) in the hippocampus of rats, which attenuates various signals transduction pathways of synaptic plasticity and consequently learning and memory deficits [56]. The depolarization of membrane above threshold causes generation of action potential which increases the level of the second messenger and activates the MAPK and mTOR pathway [57]. In addition to these, several other protein kinases and signaling pathways are involved in the induction and maintenance of synaptic plasticity. Therefore, it is important to understand the effects of Al on different signaling pathways involved in long-term memory formation.

Effect of Aluminium on Ca^{+2}-CaMKII Signaling Pathway

After the depolarization of the neuron membrane the elevated Ca^{+2} concentration in the cell activates different Ca^{+2} sensitive enzymes. Among these enzymes, the CaMKII and calcineurin (PP2B) are essential for NMDAR-dependent induction of LTP and LTD respectively [57]. The CaMKII, due to its central role in synaptic plasticity induction, is highly expressed in excitatory neurons and makes up to 1% of total proteins in the forebrain and 2% of total proteins in the hippocampus [57]. Due to binding of an inhibitory domain to the substrate binding site, the CaMKII remains inactive. After glutamatergic receptor-dependent increase in Ca^{+2} concentration in the cell, the CaMKII is activated *via* auto-phosphorylation [57]. The activated CAMKII acts on various targets that are important to orchestrate the early events of synaptic plasticity. The ability of Al to prevent Ca^{+2} influx, *via* competitively binding to Ca^{+2} channel during the period of rapid Ca^{+2} inflow [58], might cause inhibition of CaMKII activation (Fig. **1**). Wang *et al.* reported that Al causes reduction in the expression of CaMKII in the mouse brain and thus causes disruption of the CaMKII-dependent signaling pathway [59].

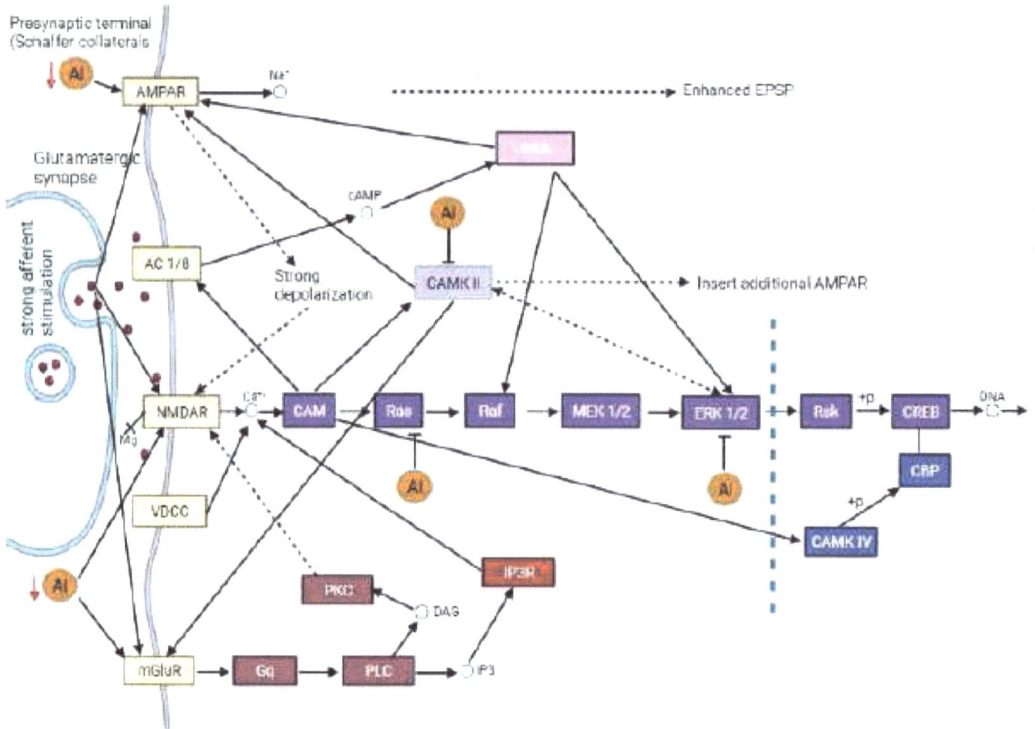

Fig. (1). Effect of Aluminium on the pathways involved in synaptic plasticity. The figure is modified from KEGG pathway and illustration created with BioRender.com. The red arrow depicts Al induced downregulation of receptor expression. The solid arrows depict direct relation while the dotted arrows show indirect interaction. The hollow circles depict chemical compounds.

Abbreviations: AMPAR, α-amino-3-hydroxy-5-methyl-4-isoxazole propionic acid receptor; AC, Adenylate cyclase; NMDAR, N-methyl-D-aspartate receptor; VDCC, voltage-dependent calcium channels; mGluR, metabotropic glutamate receptor; cAMP, cyclic adenosine monophosphate; PKA, Protein kinase A; CaMK, Ca^{2+}/calmodulin-dependent protein kinase; Ras, Reticular activating system; Raf, Rapidly Accelerated Fibrosarcoma; MEK, Mitogen-activated protein kinase; ERK, extracellular signal-regulated kinase; Rsk, Ribosomal S6 Kinase; CREB, cAMP response element-binding protein; CBP, CREB-binding protein; Gq, guanine alpha subunit; PLC, phospholipase C; PKC, Protein kinase C; IP3, inositol 1,4,5-trisphosphate; IP3R, inositol 1,4,5-trisphosphate receptor.

Effect of Aluminium on Glutamate-nitric Oxide-cyclic GMP Signaling Pathway

The elevated Ca^{+2} concentration inside the cell activates the nitric oxide synthase enzyme and formation of nitric oxide (NO) is catalyzed. This NO in turn activates guanylate cyclase which leads to the production of cyclic GMP. The cyclic GMP is a second messenger and it activates various signaling pathways [16] (Fig. **2**). Cucarella *et al*. reported the toxic effects of Al on nitric oxide-cyclic GMP (NO-cGMP) pathway about two decades ago [60]. Cucarella *et al*. reported that long-term exposure of cultured neurons with Al (for 14 days) reduced glutamate-induced activation of nitric oxide synthase by 38% and formation of cGMP by 77% [60]. Similarly, in the neuron culture of the rats prenatally exposed to Al, but not during culture, caused an 81% decrease in glutamate-induced cGMP formation [60]. Another report from the same group revealed that prenatal Al exposure reduced nitric oxide synthase and guanylate synthase content and this effect was observed to be specific for NO-cGMP pathway as the content of most other kinases remained unaffected [61]. Canales *et al*. also reported a significant decrease in the glutamate-induced cGMP production *in vitro* (cultured cerebellar neurons) and in vivo (brain microdialysis of freely moving rats [62]. This shows that impairment of the NO-cGMP pathway may contribute to Al-induced synaptic plasticity deficits. The involvement of NO pathway is also evident from the observation that L-arginin injection antagonizes the Al induced deficit in LTP. This antagonism is modulated by L-arginin-NO pathway [63].

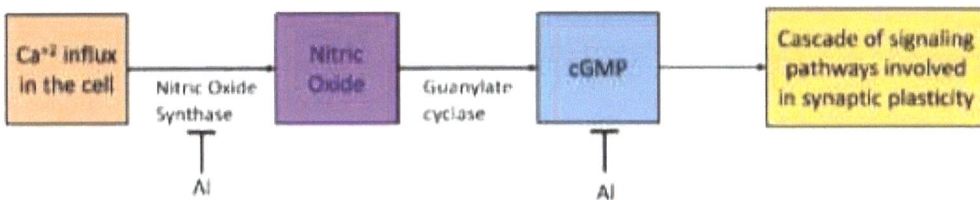

Fig. (2). Effect of Aluminium on NO-cGMP pathway.

Abbreviations: cGMP, cyclic guanine monophosphate

Effect of Aluminium on BDNF Modulated Signaling Pathways

The role of brain-derived neurotrophic factor (BDNF) in synaptic plasticity was reported more than two decades ago. The main signaling pathways mediated by BDNF include MAPK, phospholipase Cγ (PLC) and phosphatidylinositol-3 kinase (PI3K) [64]. The exposure of rats to Al resulted in the inhibition of PLC expression [39]. Al is also known to reduce the expression of BDNF *via* inhibition of plant homeodomain finger protein 8 (PHF8) [65] (Fig. **3**). Due to its inhibition, PHF8 cannot act as demethylase which in turn blocks histone H3K9 demethylation that causes reduced BDNF protein expression [66]. The BDNF gene expression is not affected by Al exposure, which shows that only the expression at translation level is hampered due to the inhibition of proteins involved in synthesis of BDNF [66].

The BDNF is involved in the synthesis of another important protein, Arc, which has a vital role in the stabilization of synaptic plasticity. In a study on SH-SY5Y human neuroblastoma cells, it was observed that pretreatment of Al(mal)$_3$ leads to a significant reduction in BDNF-induced expression of Arc. This in turn, leads to an interruption in ERK signaling pathway and LTP stabilization [67].

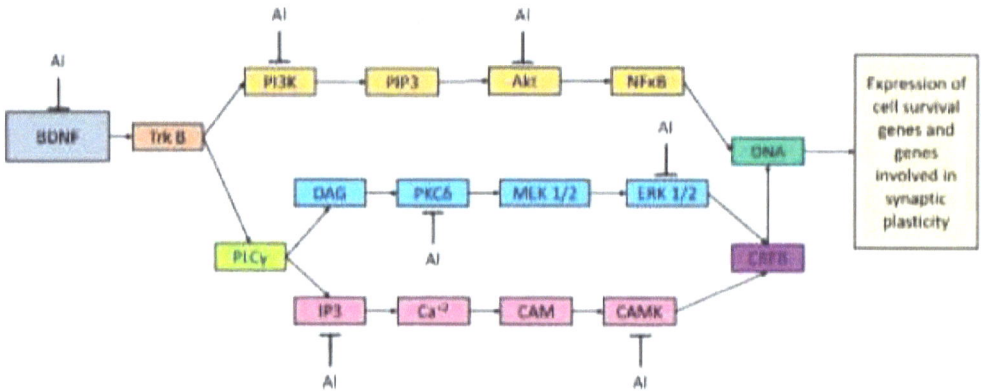

Fig. (3). Effect of Aluminium on BDNF signaling pathway.

The LTP is dependent on AMPA receptor externalization. The AMPA receptor delivery is mediated by Akt and GSK-3β signaling pathways. Moreover, cleavage of Akt *via* caspase-3 also mediates synaptic plasticity. It has been observed that Al causes a gradual decrease in Akt and GSK-3β which causes impairment in synaptic plasticity [68].

The PI3K, in conjunction with Akt and mTOR (PI3K-Akt-mTOR signaling pathway), can activate different substrates and plays an important role in cell survival and maintenance of synaptic plasticity [69]. Exposure to Al is reported to reduce the expression of Akt, PI3K and mTOR in the hippocampus of Al-exposed rats. Thus damage to this system might be involved in the Al-induced impairment of synaptic plasticity and consequently learning and memory deficits [18]. Similarly, reduced expression of PI3K, Akt and mTOR was observed in individuals having occupational exposure to Al [70]. The expression of RAS, which is another important protein in the PI3K pathway, is also reported to be reduced in the hippocampus of Al-exposed rats and might also contribute to Al-induced inhibition of LTP [69]. The subchronic exposure to 45μmol/kg Al(mal)$_3$ to wistar rats for eight weeks resulted in a significant reduction in RAS and ERK protein. The reduced level of RAS and ERK might be related to suppression of RAS-MAPK pathway that leads to impaired synaptic plasticity [71].

Effect of Aluminium on PLC Signaling Pathway

The activation of phospholipase system (PLC) through muscarinic receptors catalyzes phosphoinositide 4,5-biphosphate (PIP2) formation, *via* G protein coupling, which in turn is cleaved to inositol triphosphate (IP3) and DAG. The IP3, being soluble in water, is released into the cytoplasm where it causes release of Ca^{+2} and regulated Ca^{+2}/CaMK pathway and DAG activates membrane protein kinase C (PKC) [16]. Exposure to Al is known to inhibit PIP2 hydrolysis *via* phosphoinositide specific phospholipase C [72]. As a result, a lesser amount of IP3 is available for activation of PKC. Thus the Ca^{+2} homeostasis is affected, which in turn leads to impairment of synaptic plasticity [73]. Al is also known to inhibit the expression of PKC [56]. Thus, as a result the PLC signaling pathway is impaired and this causes damage to synaptic plasticity (Fig. **3**).

Abbreviations: BDNF, brain derived neurotropic factor; TrkB, Tropomyosin receptor kinase B; PI3K, Phosphoinositide 3-kinase; PIP3, Phosphatidylinositol-3, 4, 5-trisphosphate; Akt, Protein kinase B; NFκB, nuclear factor kappa-light-chan-n-enhancer of activated B cells; PLCγ, phospholipase Cγ; DAG, Diacylglycerol; IP3, inositol 1,4,5-trisphosphate; PKC, Protein kinase C; MEK, Mitogen-activated protein kinase; ERK, extracellular signal-regulated kinase; CaMK, Ca$^{2+/}$calmodulin-dependent protein kinase; CREB, cAMP response element-binding protein.

CONCLUDING REMARKS

The evidence indicates that many pathways are potentially involved in the Al-induced impairment in synaptic plasticity and ultimately its effects on learning and memory. However further research is required to specifically identify the effectors and regulators of these signaling pathways to better understand the simultaneous involvement of various pathways in pathogenesis of a single element.

CONSENT FOR PUBLICATION

Not Applicable.

CONFLICT OF INTEREST

The author declares no conflict of interest, financial or otherwise.

ACKNOWLEDGEMENTS

Declared none.

LIST OF ABBREVIATIONS

Al	Aluminium
LTP	Long Term Potentiation
LTD	Long Term Depression
TEA	Tetra-Ethyl Ammonium
NMDAR	N-methyl-D-aspartate receptor
VDCC	Voltage-dependent calcium channels
mGluR	Metabotropic glutamate receptor
VDAC	Mitochondrial voltage-gated channel
I$_{HVA}$	High voltage activated Ca^{+2} channels
AMPAR	α-amino-3-hydroxy-5-methyl-4-isoxazole propionic acid receptor
PI-PLC	Phosphotidylinositol-specific phospholipase C
PKC	Protein kinase C
GABA	Gamma Amino Butyric Acid
MAPK	Mitogen-activated protein kinase
mTOR	Mammalian/mechanistic target of rapamycin
PP2B	Calcineurin
NO	Nitric Oxide
NO-cGMP	Nitric oxide-cyclic GMP
PHF8	Plant homeodomain finger protein 8
Arc	Activity-regulated cytoskeleton-associated protein
Akt	Protein kinase B
GSK-3	Glycogen synthase kinase 3
PIP2	Phosphoinositide 4,5-biphosphate
AC	Adenylate cyclase
cAMP	Cyclic adenosine monophosphate
PKA	Protein kinase A
CaMK	Ca^{2+}/calmodulin-dependent protein kinase
Ras	Reticular activating system

Raf Rapidly	Accelerated Fibrosarcoma
MEK	Mitogen-activated protein kinase
Rsk	Ribosomal S6 Kinase
CREB	cAMP response element-binding protein
CBP	CREB-binding protein
Gq	Guanine alpha subunit
PLC	Phospholipase C
IP3	Inositol 1,4,5-trisphosphateinositol 1,4,5-trisphosphate
IP3R	Inositol 1,4,5-trisphosphate receptor
cGMP	Cyclic guanine monophosphate
BDNF	Brain derived neurotropic factor
TrkB	Tropomyosin receptor kinase B
PI3K	Phosphoinositide 3-kinase
PIP3	Phosphatidylinositol-3,4,5-trisphosphate
NFκB	Nuclear factor kappa-light-chain-enhancer of activated B cells
DAG	Diacylglycerol
ERK	Extracellular signal-regulated kinase

REFERENCES

[1]   Bondy SC. The neurotoxicity of environmental Aluminium is still an issue. Neurotoxicol 2010; 31(5): 575-81.
[http://dx.doi.org/10.1016/j.neuro.2010.05.009] [PMID: 20553758]

[2]   Lu X. Occupational Exposure to Aluminium and Cognitive Impairment.Neurotoxicity of Aluminium. Springer 2018; pp. 85-97.
[http://dx.doi.org/10.1007/978-981-13-1370-7_5]

[3]   Bitra VR, Rapaka D, Mathala N, Akula A. Effect of wheat grass powder on Aluminium induced Alzheimer's disease in Wistar rats. Asian Pac J Trop Med 2014; 7S1: S278-81.
[http://dx.doi.org/10.1016/S1995-7645(14)60246-7] [PMID: 25312136]

[4]   Yokel RA, McNamara PJ. Aluminium toxicokinetics: an updated minireview. Pharmacol Toxicol 2001; 88(4): 159-67.
[http://dx.doi.org/10.1034/j.1600-0773.2001.d01-98.x] [PMID: 11322172]

[5]   Ferreira PC, Piai KdeA, Takayanagui AM, Segura-Muñoz SI. Aluminium as a risk factor for Alzheimer's disease. Rev Lat Am Enfermagem 2008; 16(1): 151-7.
[http://dx.doi.org/10.1590/S0104-11692008000100023] [PMID: 18392545]

[6]   Walton JR. A longitudinal study of rats chronically exposed to Aluminium at human dietary levels. Neurosci Lett 2007; 412(1): 29-33.
[http://dx.doi.org/10.1016/j.neulet.2006.08.093] [PMID: 17156917]

[7]   Gonçalves PP, Silva VS. Does neurotransmission impairment accompany aluminium neurotoxicity? J Inorg Biochem 2007; 101(9): 1291-338.
[http://dx.doi.org/10.1016/j.jinorgbio.2007.06.002] [PMID: 17675244]

[8]   Djebli N, Rebai O. Neurotoxic effect of aluminium exploratory behaviors and spatial learning in mice.

Mov Disord 2010; 25: S400.

[9]   Exley C, House ER. Aluminium in the human brain. Monatshefte für Chemie-Chemical Monthly 2011; 142(4): 357-63.
[http://dx.doi.org/10.1007/s00706-010-0417-y]

[10]   Fulgenzi A, Vietti D, Ferrero ME. Aluminium involvement in neurotoxicity. BioMed Res Int 2014.
[http://dx.doi.org/10.1155/2014/758323]

[11]   Mardini J, Lavergne V, Ghannoum M. Aluminium transfer during dialysis: a systematic review. Int Urol Nephrol 2014; 46(7): 1361-5.
[http://dx.doi.org/10.1007/s11255-014-0752-8] [PMID: 24938693]

[12]   Campdelacreu J. Parkinson disease and Alzheimer disease: environmental risk factors. Neurologia 2014; 29(9): 541-9. [English Edition].
[http://dx.doi.org/10.1016/j.nrl.2012.04.001] [PMID: 22703631]

[13]   Inan-Eroglu E, Ayaz A. Is Aluminium exposure a risk factor for neurological disorders? J Res Med Sci 2018; 23: 51.
[http://dx.doi.org/10.4103/jrms.JRMS_921_17] [PMID: 30057635]

[14]   Xu Y, Zhang H, Pan B, Zhang S, Wang S, Niu Q. Transcriptome-wide identification of differentially expressed genes and long non-coding RNAs in Aluminium-treated rat hippocampus. Neurotox Res 2018; 34(2): 220-32.
[http://dx.doi.org/10.1007/s12640-018-9879-1] [PMID: 29460113]

[15]   Verma S, Ranawat P, Sharma N, Nehru B. Ginkgo biloba attenuates Aluminium lactate-induced neurotoxicity in reproductive senescent female rats: behavioral, biochemical, and histopathological study. Environ Sci Pollut Res Int 2019; 26(26): 27148-67.
[http://dx.doi.org/10.1007/s11356-019-05743-5] [PMID: 31321719]

[16]   Zhang H. Aluminium-Induced Electrophysiological Variation, Synaptic Plasticity Impairment, and Related Mechanism Neurotoxicity of Aluminium. Springer 2018; pp. 161-72.

[17]   Liang RF, Li WQ, Wang XH, *et al.* Aluminium-maltolate-induced impairment of learning, memory and hippocampal long-term potentiation in rats. Ind Health 2012; 50(5): 428-36.
[http://dx.doi.org/10.2486/indhealth.MS1330] [PMID: 22878356]

[18]   Li H, Xue X, Li L, *et al.* Aluminium-Induced Synaptic Plasticity Impairment *via* PI3K-Akt-mTOR Signaling Pathway. Neurotox Res 2020; 37(4): 996-1008.
[http://dx.doi.org/10.1007/s12640-020-00165-5] [PMID: 31970651]

[19]   Citri A, Malenka RC. Synaptic plasticity: multiple forms, functions, and mechanisms. Neuropsychopharmacology 2008; 33(1): 18-41.
[http://dx.doi.org/10.1038/sj.npp.1301559] [PMID: 17728696]

[20]   Farhat SM, Mahboob A, Iqbal G, Ahmed T. Aluminium-induced cholinergic deficits in different brain parts and its implications on sociability and cognitive functions in mice. Biol Trace Elem Res 2017; 177(1): 115-21.
[http://dx.doi.org/10.1007/s12011-016-0856-3] [PMID: 27709498]

[21]   Hashmi AN, Yaqinuddin A, Ahmed T. Pharmacological effects of Ibuprofen on learning and memory, muscarinic receptors gene expression and APP isoforms level in pre-frontal cortex of AlCl$_3$-induced toxicity mouse model. Int J Neurosci 2015; 125(4): 277-87.
[http://dx.doi.org/10.3109/00207454.2014.922972] [PMID: 24825584]

[22]   Mahboob A, Farhat SM, Iqbal G, *et al.* Alpha-lipoic acid-mediated activation of muscarinic receptors improves hippocampus- and amygdala-dependent memory. Brain Res Bull 2016; 122: 19-28.
[http://dx.doi.org/10.1016/j.brainresbull.2016.02.014] [PMID: 26912408]

[23]   Platt B, Carpenter DO, Büsselberg D, Reymann KG, Riedel G. Aluminium impairs hippocampal long-term potentiation in rats *in vitro* and in vivo. Exp Neurol 1995; 134(1): 73-86.
[http://dx.doi.org/10.1006/exnr.1995.1038] [PMID: 7672040]

[24]   Platt B, Reymann KG. Inhibition of TEA-induced LTP by Aluminium. Exp Neurol 1996; 141(2): 240-7.
[http://dx.doi.org/10.1006/exnr.1996.0158] [PMID: 8812157]

[25]   Qin X, Li L, Nie X, Niu Q. Effects of Chronic Aluminium Lactate Exposure on Neuronal Apoptosis and Hippocampal Synaptic Plasticity in Rats. Biol Trace Elem Res 2020; 197(2): 571-9.
[http://dx.doi.org/10.1007/s12011-019-02007-8] [PMID: 31845204]

[26]   Zhang L, Jin C, Liu Q, *et al.* Effects of subchronic Aluminium exposure on spatial memory, ultrastructure and L-LTP of hippocampus in rats. J Toxicol Sci 2013; 38(2): 255-68.
[http://dx.doi.org/10.2131/jts.38.255] [PMID: 23535404]

[27]   Chen J, Wang M, Ruan D, She J. Early chronic aluminium exposure impairs long-term potentiation and depression to the rat dentate gyrus *in vivo.* Neuroscience 2002; 112(4): 879-87.
[http://dx.doi.org/10.1016/S0306-4522(02)00138-0] [PMID: 12088747]

[28]   Wang M, Chen J-T, Ruan D-Y, Xu Y-Z. The influence of developmental period of Aluminium exposure on synaptic plasticity in the adult rat dentate gyrus *in vivo.* Neuroscience 2002; 113(2): 411-9.
[http://dx.doi.org/10.1016/S0306-4522(02)00193-8] [PMID: 12127098]

[29]   Bear MF, Connors BW, Paradiso MA. Neuroscience: past, present, and future Neuroscience: Exploring the Brain. 3rd ed. Lippincott Williams & Wilkins 2007; p. 19.

[30]   Koenig ML, Jope RS. Aluminium inhibits the fast phase of voltage-dependent calcium influx into synaptosomes. J Neurochem 1987; 49(1): 316-20.
[http://dx.doi.org/10.1111/j.1471-4159.1987.tb03432.x] [PMID: 2438384]

[31]   Platt B, Büsselberg D. Actions of Aluminium on voltage-activated calcium channel currents. Cell Mol Neurobiol 1994; 14(6): 819-29.
[http://dx.doi.org/10.1007/BF02088687] [PMID: 7641239]

[32]   Levy M, Faas GC, Saggau P, Craigen WJ, Sweatt JD. Mitochondrial regulation of synaptic plasticity in the hippocampus. J Biol Chem 2003; 278(20): 17727-34.
[http://dx.doi.org/10.1074/jbc.M212878200] [PMID: 12604600]

[33]   Dill ET, Holden MJ, Colombini M. Voltage gating in VDAC is markedly inhibited by micromolar quantities of Aluminium. J Membr Biol 1987; 99(3): 187-96.
[http://dx.doi.org/10.1007/BF01995699] [PMID: 2447281]

[34]   Chen L, Liu CJ, Tang M, *et al.* Action of Aluminium on high voltage-dependent calcium current and its modulation by ginkgolide B. Acta Pharmacol Sin 2005; 26(5): 539-45.
[http://dx.doi.org/10.1111/j.1745-7254.2005.00073.x] [PMID: 15842770]

[35]   Nday CM, Drever BD, Salifoglou T, Platt B. Aluminium interferes with hippocampal calcium signaling in a species-specific manner. J Inorg Biochem 2010; 104(9): 919-27.
[http://dx.doi.org/10.1016/j.jinorgbio.2010.04.010] [PMID: 20510457]

[36]   Büsselberg D, Platt B, Haas HL, Carpenter DO. Voltage gated calcium channel currents of rat dorsal root ganglion (DRG) cells are blocked by Al^{3+}. Brain Res 1993; 622(1-2): 163-8.
[http://dx.doi.org/10.1016/0006-8993(93)90815-5] [PMID: 8242354]

[37]   Kawahara M, Kato-Negishi M. Link between Aluminium and the pathogenesis of Alzheimer's disease: the integration of the Aluminium and amyloid cascade hypotheses. Int J Alzheimers Dis 2011; 2011: 276393.
[http://dx.doi.org/10.4061/2011/276393] [PMID: 21423554]

[38]   Iglesias-González J, Sánchez-Iglesias S, Beiras-Iglesias A, Méndez-Álvarez E, Soto-Otero R. Effects of aluminium on rat brain mitochondria bioenergetics: an *in vitro* and in vivo study. Mol Neurobiol 2017; 54(1): 563-70.
[http://dx.doi.org/10.1007/s12035-015-9650-z] [PMID: 26742531]

[39]   Jin C, Wu S, Zhou P, *et al.* Effect of Aluminium on Ca²+ concentration and expression of phospholipase C and NMDA receptor α genes in hippocampus of weaning rats as well as their neural behavior through subchronic exposure. Zhonghua lao dong wei sheng zhi ye bing za zhi = Chinese J Indust hygiene Occup Dis 28(9): 648-51.2010;

[40]   Malenka RC, Bear MF. LTP and LTD: an embarrassment of riches. Neuron 2004; 44(1): 5-21. [http://dx.doi.org/10.1016/j.neuron.2004.09.012] [PMID: 15450156]

[41]   Xiao H, Liu B, Chen Y, Zhang J. Learning, memory and synaptic plasticity in hippocampus in rats exposed to sevoflurane. Int J Dev Neurosci 2016; 48: 38-49. [http://dx.doi.org/10.1016/j.ijdevneu.2015.11.001] [PMID: 26612208]

[42]   Miyamoto E. Molecular mechanism of neuronal plasticity: induction and maintenance of long-term potentiation in the hippocampus. J Pharmacol Sci 2006; 100(5): 433-42. [http://dx.doi.org/10.1254/jphs.CPJ06007X] [PMID: 16799259]

[43]   Platt B, Haas H, Büsselberg D. Aluminium reduces glutamate-activated currents of rat hippocampal neurones. Neuroreport 1994; 5(17): 2329-32. [http://dx.doi.org/10.1097/00001756-199411000-00030] [PMID: 7533557]

[44]   Ren P, Kang P, Li Z, Zhang H, Niu Q. [Impact of chronic Aluminium exposure on NMDAR1 in the cortex and peripheral blood lymphocytes in rats]. Wei Sheng Yen Chiu 2017; 46(1): 15-20. [PMID: 29903145]

[45]   Yuan C-Y, Hsu G-SW, Lee Y-J. Aluminium alters NMDA receptor 1A and 2A/B expression on neonatal hippocampal neurons in rats. J Biomed Sci 2011; 18(1): 81. [http://dx.doi.org/10.1186/1423-0127-18-81] [PMID: 22067101]

[46]   Nam SM, Yoo DY, Kwon HJ, *et al.* Effects of long-term exposure to Aluminium in the hippocampus in the type 2 diabetes model rats. Toxicol Res (Camb) 2018; 8(2): 206-15. [http://dx.doi.org/10.1039/C8TX00192H] [PMID: 30931101]

[47]   Jin C, Wu S, Lu X, *et al.* The expression changes of N-methyl-D-aspartic acid receptor in the hippocampus of offspring from female rats exposed to Aluminium in the pregnancy and lactation. Zhonghua lao Dong wei Sheng zhi ye Bing za zhi = Chinese J Indust Hygiene and Occup Dis 2011; 29(3): 198-201.

[48]   Zhi-jun Y, Wei X. Effect of Aluminium Exposure on NMDA Receptors of Rat Hippocampus [J]. Progress of Anatomical Sciences 2008; 4

[49]   Pan B, Li Y, Zhang J, *et al.* Role of mGluR 1 in synaptic plasticity impairment induced by maltol aluminium in rats. Environ Toxicol Pharmacol 2020; 78: 103406. [http://dx.doi.org/10.1016/j.etap.2020.103406] [PMID: 32438325]

[50]   Song J, Liu Y, Zhang HF, Zhang QL, Niu Q. Effects of exposure to Aluminium on long-term potentiation and AMPA receptor subunits in rats in vivo. Biomed Environ Sci 2014; 27(2): 77-84. [PMID: 24625397]

[51]   SONG J, NIU Q Effect of subchronic Aluminium exposure on learning and memory abilities and the expression of AMPA receptor gene in rats. China Occupational Medicine 2013; p. 2. J

[52]   Barberis A, Bacci A. Editorial: Plasticity of GABAergic synapses. Front Cell Neurosci 2015; 9: 262. [http://dx.doi.org/10.3389/fncel.2015.00262] [PMID: 26217186]

[53]   Vashchinkina E, Panhelainen A, Aitta-Aho T, Korpi ER. GABAA receptor drugs and neuronal plasticity in reward and aversion: focus on the ventral tegmental area. Front Pharmacol 2014; 5: 256. [http://dx.doi.org/10.3389/fphar.2014.00256] [PMID: 25505414]

[54]   Trombley PQ. Selective modulation of GABAA receptors by Aluminium. J Neurophysiol 1998; 80(2): 755-61. [http://dx.doi.org/10.1152/jn.1998.80.2.755] [PMID: 9705466]

[55]   L Blaylock R. Aluminium induced immunoexcitotoxicity in neurodevelopmental and

neurodegenerative disorders. Curr Inorg Chem 2012; 2(1): 46-53.
[http://dx.doi.org/10.2174/1877944111202010046]

[56] Wang B, Xing W, Zhao Y, Deng X. Effects of chronic Aluminium exposure on memory through multiple signal transduction pathways. Environ Toxicol Pharmacol 2010; 29(3): 308-13.
[http://dx.doi.org/10.1016/j.etap.2010.03.007] [PMID: 21787618]

[57] Kennedy MB. Synaptic signaling in learning and memory. Cold Spring Harb Perspect Biol 2013; 8(2): a016824.
[http://dx.doi.org/10.1101/cshperspect.a016824] [PMID: 24379319]

[58] Moraes TP, Bucharles SG, Ribeiro SC, Frumento R, Riella MC, Pecoits-Filho R. Low-calcium peritoneal dialysis solution is effective in bringing PTH levels to the range recommended by current guidelines in patients with PTH levels < 150 pg/dL. J Bras Nefrol 2010; 32(3): 275-80.
[http://dx.doi.org/10.1590/S0101-28002010000300009] [PMID: 21103691]

[59] Wang B, Zhao J, Yu M, *et al.* Disturbance of intracellular calcium homeostasis and CaMKII/CREB signaling is associated with learning and memory impairments induced by chronic Aluminium exposure. Neurotox Res 2014; 26(1): 52-63.
[http://dx.doi.org/10.1007/s12640-013-9451-y] [PMID: 24366850]

[60] Cucarella C, Montoliu C, Hermenegildo C, *et al.* Chronic exposure to Aluminium impairs neuronal glutamate-nitric oxide-cyclic GMP pathway. J Neurochem 1998; 70(4): 1609-14.
[http://dx.doi.org/10.1046/j.1471-4159.1998.70041609.x] [PMID: 9580158]

[61] Llansola M, Miñana MD, Montoliu C, *et al.* Prenatal exposure to Aluminium reduces expression of neuronal nitric oxide synthase and of soluble guanylate cyclase and impairs glutamatergic neurotransmission in rat cerebellum. J Neurochem 1999; 73(2): 712-8.
[http://dx.doi.org/10.1046/j.1471-4159.1999.0730712.x] [PMID: 10428068]

[62] Canales JJ, Corbalán R, Montoliu C, *et al.* Aluminium impairs the glutamate-nitric oxide-cGMP pathway in cultured neurons and in rat brain in vivo: molecular mechanisms and implications for neuropathology. J Inorg Biochem 2001; 87(1-2): 63-9.
[http://dx.doi.org/10.1016/S0162-0134(01)00316-6] [PMID: 11709215]

[63] Zou B, Zhang Z, Xiao H, Li A. Effect of Aluminium on long-term potentiation and its relation to L-arg-NO-pathway in hippocampal CA3 area of rats. J Tongji Med Univ 1998; 18(4): 193-6.
[http://dx.doi.org/10.1007/BF02886470] [PMID: 10806843]

[64] Gottschalk WA, Jiang H, Tartaglia N, Feng L, Figurov A, Lu B. Signaling mechanisms mediating BDNF modulation of synaptic plasticity in the hippocampus. Learn Mem 1999; 6(3): 243-56.
[PMID: 10492006]

[65] Li Z, Kang P, Zhang Hu, Nie X, Yuan Y, Niu Q. [The pilot study on the expression of PHF8, H3K9me2, BDNF and LTP in the hippocampus of rats exposed to Aluminium]. Zhonghua Lao Dong Wei Sheng Zhi Ye Bing Za Zhi 2016; 34(1): 18-22.
[PMID: 27014810]

[66] Li H, Xue X, Li Z, Pan B, Hao Y, Niu Q. Aluminium-induced synaptic plasticity injury *via* the PHF8-H3K9me2-BDNF signalling pathway. Chemosphere 2020; 244: 125445.
[http://dx.doi.org/10.1016/j.chemosphere.2019.125445] [PMID: 31835052]

[67] Chen T-J, Cheng H-M, Wang D-C, Hung H-S. Nonlethal Aluminium maltolate can reduce brain-derived neurotrophic factor-induced Arc expression through interrupting the ERK signaling in SH-SY5Y neuroblastoma cells. Toxicol Lett 2011; 200(1-2): 67-76.
[http://dx.doi.org/10.1016/j.toxlet.2010.10.016] [PMID: 21040763]

[68] Zhang H, Yang X, Qin X, Niu Q. Caspase-3 is involved in Aluminium-induced impairment of long-term potentiation in rats through the Akt/GSK-3β pathway. Neurotox Res 2016; 29(4): 484-94.
[http://dx.doi.org/10.1007/s12640-016-9597-5] [PMID: 26787483]

[69] Song J, Liu Y, Zhang HF, Niu Q. The RAS/PI3K pathway is involved in the impairment of long-term

potentiation induced by acute Aluminium treatment in rats. Biomed Environ Sci 2016; 29(11): 782-9.
[PMID: 27998384]

[70]   Shang N, Zhang P, Wang S, *et al.* Aluminium-Induced Cognitive Impairment and PI3K/Akt/mTOR
         Signaling Pathway Involvement in Occupational Aluminium Workers. Neurotox Res 2020; 38(2):
         344-58.
         [http://dx.doi.org/10.1007/s12640-020-00230-z] [PMID: 32506341]

[71]   Song J, Li Z, Zhang L, Niu Q. . Effects of subchronic Aluminium exposure on long-term potentiation
         and activities of RAS and extracellular regulated protein kinases in rats. Zhonghua lao Dong wei
         Sheng zhi ye Bing za zhi Chinese J Indust Hygiene Occup Dis 2017; 35(5): 328-1.

[72]   Nostrandt AC, Shafer TJ, Mundy WR, Padilla S. Inhibition of rat brain phosphatidylinositol-specific
         phospholipase C by Aluminium: regional differences, interactions with Aluminium salts, and
         mechanisms. Toxicol Appl Pharmacol 1996; 136(1): 118-25.
         [http://dx.doi.org/10.1006/taap.1996.0014] [PMID: 8560464]

[73]   Walton JR. Aluminium disruption of calcium homeostasis and signal transduction resembles change
         that occurs in aging and Alzheimer's disease. J Alzheimers Dis 2012; 29(2): 255-73.
         [http://dx.doi.org/10.3233/JAD-2011-111712] [PMID: 22330830]

CHAPTER 5

Aluminium and other Metals Exposure Cause Neurological Disorders: Evidence from Clinical/ human Studies

Zehra Batool[1], Laraib Liaquat[2], Tuba Sharf Batool[3], Rida Nisar[4] and Saida Haider[5,*]

[1] *Dr. Panjwani Center for Molecular Medicine and Drug Research, International Center for Chemical and Biological Sciences, University of Karachi, Karachi, Pakistan*

[2] *Multidisciplinary Research Lab, Bahria University Medical and Dental College, Bahria University, Karachi, Pakistan*

[3] *Atta-ur-Rahman School of Applied Biosciences, National University of Sciences and Technology, Islamabad, Pakistan*

[4] *HEJ Research Institute of Chemistry, International Center for Chemical and Biological Sciences, University of Karachi, Karachi, Pakistan*

[5] *Neurochemistry and Biochemical Neuropharmacology Research Unit, Department of Biochemistry, University of Karachi, Karachi, Pakistan*

Abstract: Exposure to Aluminium and other heavy metals has become a serious concern in today's modern life. Due to excessive use and improper disposal of heavy metals, the entire food chain is being contaminated, which is imposing various health risks for humans and other living organisms. These heavy metals particularly induce oxidative stress through different mechanisms which can ultimately interfere with the normal physiological activities. Brain is highly prone to oxidative stress due to its rich polyunsaturated content and high oxygen consumption than the periphery. Therefore, emphasis has been given to neurotoxicological effects produced by exposure to heavy metals. In this regard, the effects of both essential and non-essential heavy metals have been investigated in various clinical studies which are demonstrating them as a serious threat to normal brain function. This chapter summarizes the neurotoxicological effects of heavy metals which have been revealed in various human studies.

Keywords: Clinical Studies, Heavy Metals, Neurological Disorders, Oxidative Stress, Toxicity Mechanism.

* **Corresponding author Dr. Saida Haider:** Neurochemistry and Biochemical Neuropharmacology Research Unit, Department of Biochemistry, University of Karachi, Karachi, Pakistan; Tel:+92-21-99261313-17; E-mail: saida-h1@hotmail.com

Touqeer Ahmed (Ed.)

INTRODUCTION

Exposure to metals through various sources, including inhalation and ingestion, can result in accumulation in the body. Some of the metals have an essential role in the physiological and biochemical functions at an appropriate concentration. However, at higher concentrations they can accumulate in various vital organs of the body and can cause toxicity. These metals include chromium, cobalt, copper, iron, magnesium, manganese, molybdenum, nickel, selenium, and zinc. Toxic heavy metals having no known biological activities are considered hazardous metals since they can cause toxicity in the human body even at very low doses. All these metals can readily cross the blood brain barrier and can accumulate there, which may result in neurotoxicity. The mammalian nervous system becomes susceptible to metal-related redox damages as a consequence of several biochemical and physiological functions, including; high consumption of oxygen in neuronal cells as compared to the periphery, increased generation of reactive species due to reduced efficiency of mitochondria during aging, vulnerability to lipid peroxidation due to abundant content of unsaturated fats in the brain, the increased tendency of neurotransmitters to become oxidized and insufficiency of certain antioxidant mechanisms [1]. Metal-induced neurotoxicity is considered as one of the major reasons for neuronal injuries leading to neurological disorders such as Alzheimer's disease, Parkinson's disease, amyotrophic lateral sclerosis, autism spectrum disorders, Huntington's disease, multiple sclerosis, Wilson's disease, Guillain–Barré disease, and Gulf War syndrome. There are various mechanistic pathways through which metal can induce neurotoxicity, such as generation of reactive oxygen and nitrogen species, production of pro-inflammatory biomolecules, suppression of antioxidants, mitochondrial dysfunction and/or imbalance of calcium homeostasis. This chapter represents the clinical studies which have been done to demonstrate the neurological disorders induced by metal toxicity.

NEUROLOGICAL DISORDERS INDUCED BY HAZARDOUS METALS ALUMINIUM

Mechanism of Aluminium -Induced Neurotoxicity

Aluminium is associated with neurodegenerative disorders and other complications, including Alzheimer's disease, Parkinson's disease, amyotropic lateral sclerosis, multiple sclerosis, autism and epilepsy and in all these diseases Aluminium serves as a toxic co-factor. Aluminium enhances inflammatory processes in brain by various mechanisms such as by activation of microglia and by inducing pro-inflammatory gene expression. Exacerbation of inflammatory processes is similar to those observed in neurodegenerative disorders such as Alzheimer's disease brain [1].

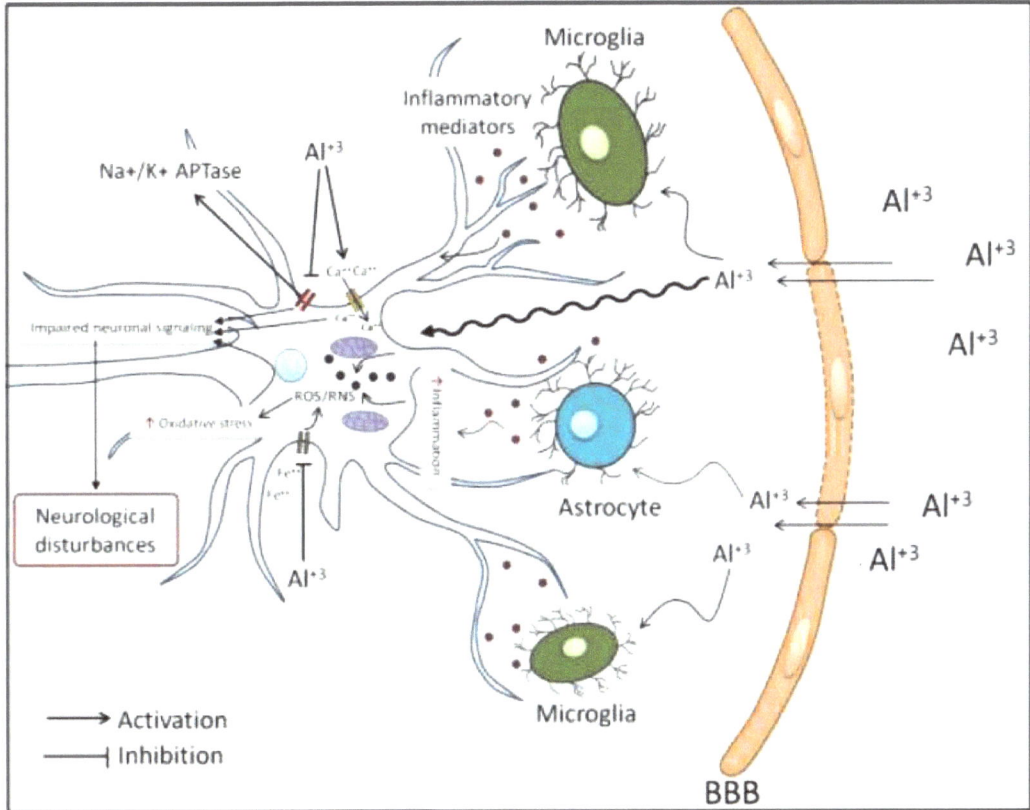

Fig. (1). Proposed mechanism of action of Aluminium-induced neurotoxicity. Refer to the text for detail. ROS: reactive oxygen species; RNS: reactive nitrogen species.

Alteration of nerve cells morphology is the most reported feature of Aluminium neurotoxicity following its exposure [2]. Permeability of blood brain barrier is compromised with age and as a result of this, substances that are confined to systemic circulation can enter the brain easily. With age, the cerebral levels of Aluminium increase and inside the brain, it expedites the aging process by oxidative alteration. As Aluminium enters the brain, it initiates the production of reactive oxygen species and activates glial cells; both of these pathways stimulate a chronic inflammatory response that eventually results in neurodegeneration [3]. Aluminium toxicity includes exacerbation of oxidative stress by increasing iron-driven and superoxide oxidation, by interfering transport and storage of iron, impairing antioxidant enzyme activities and increasing lipid peroxidation. It also disrupts calcium homeostasis, leading to a marked increase in intracellular concentration and deregulation of calcium signaling. It has been stated that Aluminium interferes with acetyl-CoA metabolism, an important precursor of acetylcholine. Aluminium leads to decreased levels of acetyl-CoA in mitochondria. Aluminium, by compromising the state of membrane polarization and magnitude of Na^+ gradient to which neurotransmitter transport is dependent, produces an inhibitory effect on (Na^+/K^+) ATPase that further leads to a decline in neurotransmitter functions. It is reported that even a submillimolar concentration of Aluminium produces inhibition of (Na^+/K^+) ATPase(Fig. 1). Monoamine oxidase B (MAO-B) is the enzyme responsible to metabolize dopamine. Marked activation of MAO-B following either acute or chronic Aluminium exposure has been reported [2].

Aluminium-Induced Neurological Disorders: Clinical Evidences

Aluminium is the most reported toxicant of biological system; studies have linked it with serious disturbances of the nervous system. Various forms of Aluminium are toxic to the nervous system and can cause major brain disorders [4]. Dialysis dementia also known as dialysis-associated encephalopathy syndrome is caused due to accumulation of intravenously administered Aluminium in dialysis fluid and observed in patients subjected to chronic dialysis due to renal failure [5]. Symptoms that are associated with dialysis dementia include verbal dysfunctions such as speech impairments (motor aphasia, dyspraxia, dysarthria, stuttering) motor dysfunctions (seizures, motor apraxia, tremors, myoclonic jerks, twitches) and cognitive and behavioral dysfunctions (progressive dementia, psychosis, confusion, paranoia) [6].

Chronic exposure of Aluminium in the form of antacids is associated with craniosynostosis (premature ossification of the skull and obliteration of the sutures) in infants [4]. Occupational exposure to Aluminium in the form of Al-oxide fumes through inhalation is reported to be associated with insomnia, headache, emotional irritability, concentration difficulties and mood liability [7]. Intravenous administration of standard infant feeding solution is another source of Aluminium exposure and is reported to be associated with reduced development attainment [8]. Potable water contamination of Aluminium-sulphate produces sporadic early-onset amyloid angiopathy along with difficulties in words, progressive dementia, cerebral ischemia and visual hallucination [9].

Aluminium is the most commonly used vaccine adjuvant. Administration of multiple vaccinations containing Aluminium adjuvant over a short period of time is associated with detrimental effects such as Gulf war syndrome [10]. Multiple sclerosis is also reported following the use of vaccines containing different compounds of Aluminium. Majority of the vaccines contain alum or either used in connection with alum containing adjuvants and increased excretion of Aluminium was also evident in those patients [10]. Parenteral exposure of Aluminium <10 days at a dose of 20 µg/kg is linked to long term damaging effects on brain development in preterm infants [8]. In 2003, US Food and Drug Association (FDA) set a limit for Aluminium exposure at no higher than 4 to 5 µg/kg bw/day for premature neonates, stating that Aluminium above this range is associated with serious damage to the central nervous system and bone loss [11].

Increased Aluminium levels have also been observed in other less common brain disorders including Hallervorden-Spatz disease and Guamanian Parkinsonian-ALS constellation. Various epidemiological studies have connected the risk of neurodegenerative disorders with Aluminium in drinking water [12]. A dose response correlation between Aluminium in drinking water and incidence of Alzheimer's disease is also found. Elderly population exposed to Aluminium in water (100µg/L) is also at risk of developing Alzheimer's disease [13]. Use of deferoxamine in Aluminium related bone disease is associated with elevated Aluminium levels in serum that later produce dementia in patients [14].

A study was conducted on a brain of 66 year old man after his death, working at a nuclear fuel and space industry, expressed repeated episodes of headaches, tiredness, memory deterioration and depression. Detailed study of the brain revealed that Aluminium sulphate dust is associated with accumulation of argyrophilic β amyloid plaques and neurofibrillary tangles profusion in all areas of cerebral cortex that lead to progression of Alzheimer's disease [15]. A 36 year multicenter study on Aluminium content in 511 samples with different brain disorders also reported Aluminium as a potential causative factor in the progression of Alzheimer's disease, dialysis dementia syndrome and Down's syndrome [16]. Clinical studies have connected elevated levels of Aluminium in the brain with the progression of Alzheimer's disease. Several studies have reported Aluminium as a strong inducer of neurological disorders. A morbidity study conducted between 1988 and 1989 revealed that workers exposed to Aluminium performed less well on cognitive assessments as compared to unexposed workers. Another study was conducted using 10-year residential history on the effects of Aluminium exposure using municipal drinking water and they found high probability of Alzheimer's disease development in areas where the amount of Aluminium was ≥ 100 µg /L in drinking water. Studies using more sensitive detections methods have further confirmed the presence of significantly elevated levels of Aluminium in inferior parietal lobe, hippocampus, medial and superior temporal gyri of Alzheimer's disease patients than control samples [17]. During an unfortunate incident happened in Camelford, England, the Aluminium level in drinking water was accidently raised to a point that produced serious cognitive deterioration in residents of that particular area [18].

ARSENIC

Mechanism of Arsenic-Induced Neurotoxicity

Electrophysiological studies performed on patients with arsenic induced neurological disorders have revealed neurotoxic nature of arsenic and various neuropathies were observed in patients, including reduced velocity of neuronal conduction and axonal degeneration [19]. Observed toxic effects of arsenic are due to inactivation of enzymes involved in cellular energy pathway. Arsenic can also react with thiol group of protein and enzyme and results in inhibition of their catalytic action. Arsenic has a role in generation of free radical species which represents the way by which arsenic induces neurotoxicity. Arsenic affects mitochondrial functions and results in disturbances of brain functions because proper functioning of brain depends largely on efficient mitochondrial functions [20]. Neuronal structural, pathological, altered synaptic structure, morphological changes, and aggregation of pathological protein such neurofibrillary tangles have been reported following arsenic-induced toxicity [21, 22]. It also affects the neurotransmitter levels in brain including dopamine, 5-HT and norepinephrine [23]. Moreover, reduced neuropsychological functions have also been reported [24] (Fig. **2**).

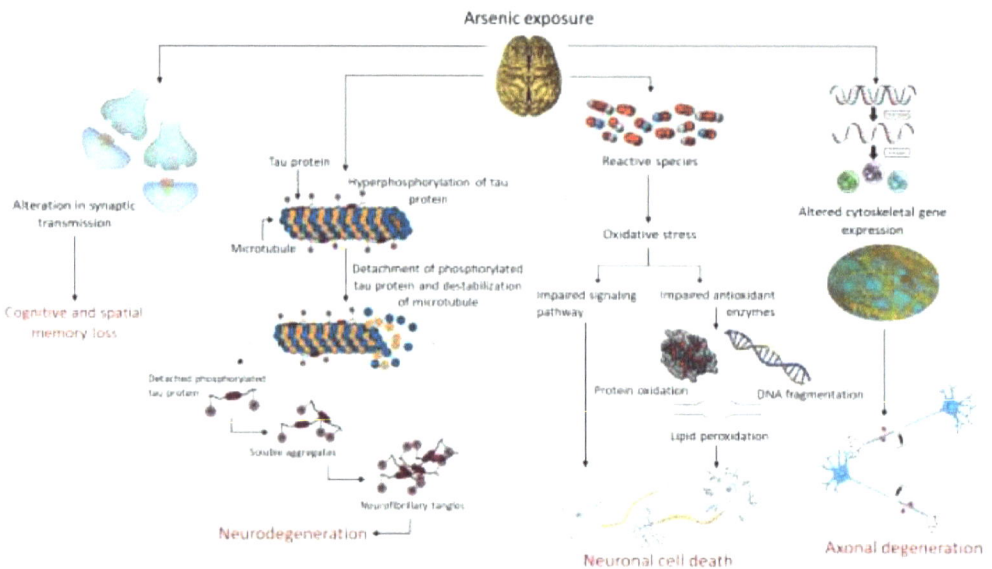

Fig. (2). Reported consequences of arsenic-induced neurotoxicity.

Arsenic-Induced Neurological Disorders: Clinical Evidences

Arsenic exists in inorganic and organic species and both forms further exist in pentavalent and trivalent oxidation states. One of the main sources of arsenic toxicity is contaminated ground water through agriculture and industrial uses [25]. The symptoms that are observed in residents exposed to arsenic contaminated drinking water include peripheral neuropathy and cognitive disturbances [26]. Data from clinical studies have also reported a close link between arsenic exposure and various neurodegenerative disorders, including autism spectrum disorder, multiple sclerosis, amyotrophic lateral sclerosis, Gulf war illness and more commonly Alzheimer's disease [27].

Exposure to arsenic metabolites is associated with severe and highly variable long-term health effects, mainly brain disorders, lung and skin cancer, cardiovascular diseases and hypertension. Brain disorders may occur within a few hours of arsenic ingestion and the symptoms may appear usually after 2-8 weeks. Neurological disorder that is associated with arsenic exposure is symmetrical sensorimotor neuropathy similar to Guillain-Barré syndrome [28]. Primarily the clinical features associated with arsenic induced neurological disorder are numbness and pain mainly in feet soles and paresthesia [29].

All forms of arsenic can cross blood brain barrier and accumulate in various region of brain and can cause serious neurological disorders [20]. Calderon and colleagues reported that arsenic influences neuropsychological development in children such as long-term memory and verbal abilities, pattern memory and switching attention was also severely affected due to exposure of arsenic [30]. Arsenic in drinking water is also associated with decreased motor functions in children [31].

Arsenic is associated with neurodegenerative mechanism in brain by inducing hyper phosphorylation and altering the cytoskeletal protein composition. Such variations lead to disruption of brain cytoskeletal framework and represent main pathogenic mechanism of arsenic induced neurotoxicity [32]. Formation of neurofibrillary tangles and amyloid B plaques as a pathogenic factor for the neurodegenerative mechanism of Alzheimer's disease has also been reported [33]. Most frequent neuropathology associated with intoxication of inorganic arsenic is distal symmetrical polyneuropathy. Depending on the dosage and duration of arsenic exposure the pathological features associated with distal symmetrical polyneuropathy are segmental demyelination and axonal degeneration. Arsenic produces such pathological changes by interfering with pyruvate oxidation [34].

Neuronal developmental process is more susceptible to the damage caused by heavy metal toxicity than mature brain [35]. In mid 1950s an unfortunate incident happened in japan due to developmental arsenic toxicity. As per officials record more than one hundred infants died from arsenic poisoning in dried milk powder

Morinaga [36]. Arsenic poisoning due to Morinaga infant dried milk powder resulted in most serious fatality rate due to food poisoning in Japan and even the six hundred surviving victims experienced serious sequelae mainly mental retardation, neurological disease and other disabilities in their 50s. Along with this occurrence of neurological disorders mainly mental retardation of brain damage, epilepsy and low intelligence quotient (IQ) was also evident in victims [37]. Possible cause of arsenic poisoning in dried milk powder was low quality disodium phosphate containing low amount of arsenic that was added to cow's milk as stabilizer [38].

MERCURY

Mechanism of Mercury-Induced Neurotoxicity

Various studies have highlighted the injurious effects of mercury (Hg) through dental exposure on brain. Studies conducted on American dental professional have highlighted the neurotoxic nature of mercury and reported that various genetic mutation are responsible for neurological outcomes due to mercuric exposure including SLC6A4, COMT, CPOX4 and BDNF genetic mutations. Such mutations are further associated with impaired cognitive flexibility, manual coordination, working memory, attention and altered mood states [39 - 43]. Neurobehavioral effects in dental workers exposed to low mercuric levels are also reported with characteristic changes in brain including microtubule deterioration, elevation of amyloid protein expression and altered neurotransmission functions such as inhibition or increase release at motor nerve endings. In comparison to Hg^{+2}, non-polar and non-oxidized Hg^0 can cross blood brain barrier and can reach the brain cells [44]. In the brain, Hg^0 is oxidized to Hg^{+2} and confined in the brain cells. Inside the brain Hg^{+2} ions produce neurotoxic effects by binding with higher affinity to essential thiol groups on enzymes and by altering the functions of neuronal cytoskeleton network including microtubules and other necessary structural components. Such circumstances result in reduced neurotransmitter release and other functions [45, 46].

Mercury-Induced Neurological Disorders: Clinical Evidences

Mercury was first recognized as highly toxic in nature in Japan when it was found as a causative factor for an epidemic situation which is now known as Minamata disease [47]. Minamata disease is a serious neurological disorder caused by mercuric poisoning and the symptoms that are associated with Minamata disease include numbness in hands and feet, ataxia and damage to sensory system including hearing, vision and speech and general muscles weakness. In Minamata, Japan, a chemical plant used an inorganic mercury compound in a chemical process as catalyst that led to the conversion of inorganic mercury compound to organomercury. After many years of this continuous practice, people of Minamata experienced some unusual symptoms associated with Hg poisoning and after systematic investigation it was revealed that organomercurial compounds were being released into sea water and concentrated in marine animals that were the main part of diet of the local population [48]. Similar situation happened in Sweden, when local population consumed fish contaminated with an organic Hg compound used as a pesticide by farmers [49].

In 1974 in Iraq, people got exposed to mercury poising by consuming rice treated with pesticide containing high amount of mercury [50]. Another incident happened in Ghana due to consumption of ethylmercury-contaminated maize by the local community, out of 144 cases of mercury poisoning 20 cases ended up in fatalities. In 1970 cases of ethylmercury poisoning was also reported in china due to consumption of ethylmercury chloride contaminated rice by farmers [51]. Mercury has been reported to induce damage to DNA, disrupts neuronal migration and inhibits mitosis [52] and is highly neurotoxic during prenatal and postnatal periods but the results are more profound on the developing embryo that later produce cerebral palsy, mental retardation, seizures and ultimately death [53]. In adults, mercury has been associated with severe neurogenic pain syndrome that requires inpatient pain management. This condition later developed into a type of severe motor neuropathy with signs and symptoms similar to Guillain-Barre-like illness and axonal degeneration [54]. Extensive literature also reported the occurrence of neurologic conditions due to mercury poisoning characterized by amyotropic lateral sclerosis-like illness [54, 55].

Dental workers get occupationally exposed to methyl mercury vapors during amalgam preparation as a filling material and also during drilling, insertion and plastering procedure [56]. Mercury releases from amalgam fillings is the primary route of exposure of inorganic and metallic mercury to human population and it is also a primary source of mercury to reach to systematic circulation and then to brain [57]. Mercury vapors can cross blood brain barrier and can further accumulate in different brain regions. Mercury preferentially accumulates in those brain regions that are involved in behavior and personality, so the toxic effects caused by mercury vapors are mostly neuropsychiatric in nature. Mainly methyl mercury is the compound that usually targets sensorimotor functions that later result in altered coordination, equilibrium and motor control. Woods and coworkers have reported that the incidence of mercuric toxicity associated neurobehavioral defects are more common in children particularly boys compared to adults because of their small body mass and more rapidly developing and fragile body systems such as nervous and metabolic system [58].

Vaccines contain significant amount of mercury in the form of Hg-containing preservative thimerosal (sodium elthylmercury thiosalicylate), along with this fish consumption also lead to increased level of Hg in the blood. Heavy metals such as mercury has been reported to be involved in the underlying pathological mechanism of neurodegenerative disorders such as Amyotrophic lateral sclerosis, Alzheimer's disease, Parkinson's disease and multiple sclerosis [59, 60]. Studies have also shown its connection with some other brain disorders such as autism spectrum disorder [61, 62]. Other neurological disorders that are reported to be caused as a consequence of mercury exposure include ataxia, paresthesia, tremors, sensory disturbances such as impairment of hearing and the symptoms usually appear in human body 150 days after exposure [63].

LEAD

Mechanism of Lead-Induced Neurotoxicity

Lead gets access to the human body through oral ingestion and inhalation. In the living body lead can bind with the red blood cells and can accumulate in the body [64]. Lead is associated with multiple pathogenic processes such as oxidative stress, disrupt mitochondrial functions, Golgi apparatus alteration and increase gliofilaments on astrocytes [65]. Lead also found to be a causative factor for imbalanced Ca^{2+} homeostasis and interfere with phosphorylation mechanism [66]. Lead is reported to cause inhibitory effects on brain growth and development and impair the behavioral and cognitive functioning. It predominantly affects neurotransmitter receptors and protein complexes such as protein kinase C and N-methyl-D-aspartate (NMDA) subtype of glutamate receptor [67]. Exposure to lead has shown to inhibit voltage-gated calcium channel and NMDA receptors leading to the inhibition of Ca++ dependent pathway, and inhibits hippocampal-dependent learning and memory. Lead also inactivates cyclic AMP-response element binding protein (CREB) and brain derived neurotrophic factor (BDNF). The neurotrophin factor BDNF is a crucial molecule involves in synaptic development and neurotransmitter release whereas the transcriptional factor CREB is involved in essential steps of memory consolidation which is dependent on the activation of NMDA receptors [68]. Therefore, exposure to lead can cause behavioral and cognitive deficits which may lead to neurodegenerative diseases (Fig. **3**).

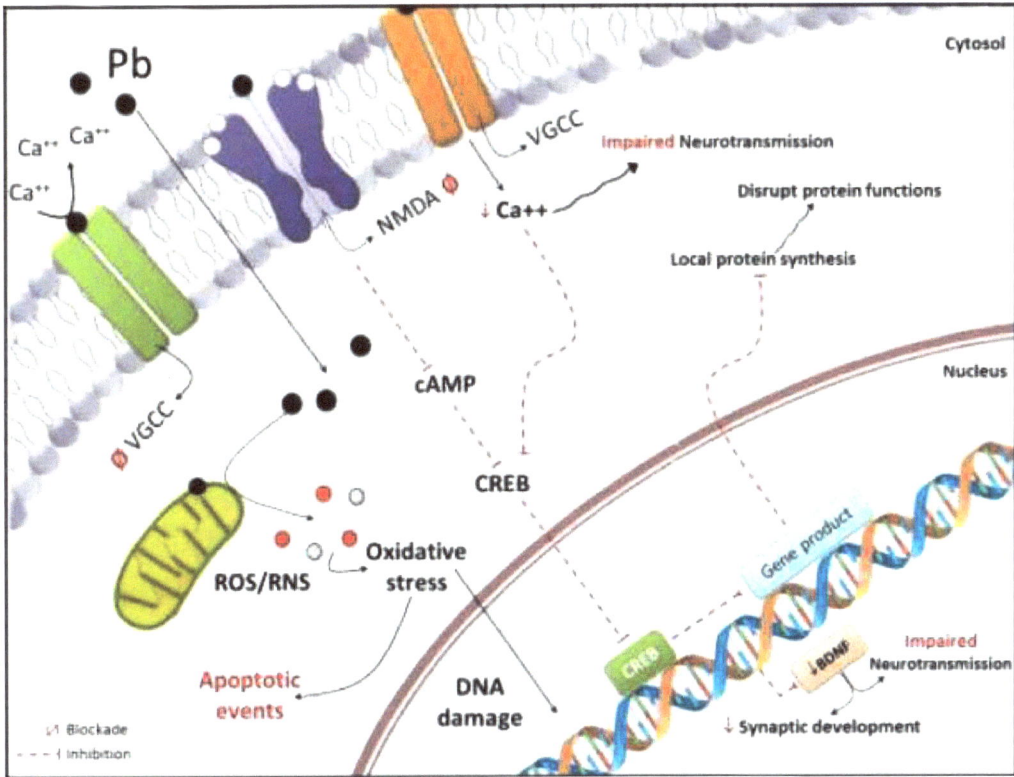

Fig. (3). Suggested mechanism of action of lead-induced neurotoxicity. Refer the text for detail. VGCC – Voltage-gated calcium channel; N-methyl-D-aspartate (NMDA) glutamate receptor.

Lead-Induced Neurological Disorders: Clinical Evidences

Lead has been recognized as a toxic metal to human body. Majority of recent investigations have highlighted the involvement of low-level lead exposure and neurobehavioral deficits mainly cognitive dysfunctions displayed from childhood through adolescence. Lead poisoning is also associated with other symptoms such as delinquency and aggressiveness. Environmental, domestic and occupational ways are the major sources of lead contamination and exposure to human body [69].

Since the ancient time, detrimental effects of lead poisoning are well known but the recent studies have highlighted the severe consequences of lead exposure on higher functions of brain. Inside the brain, hippocampus is the primary target of lead, along with other major regions of brain, lead has the ability to easily accumulate in the hippocampal region [70] and produce neurotoxic effects. Lead intoxication is associated with reduced cognitive functions, intelligence, executive functions, attention, emotion, language, memory, speed processing, visuospatial and motor skills [64]. Various studies have reported that lead exposure in children is associated with dose-dependent gradual decline in intellectual ability, impairment in verbal concept, difficulties in grammatical reasoning and inabilities in command following [71, 72].

In Germany, investigation on neurotoxic effects of low level of lead exposure in a cohort of 6-7 year old children revealed that lead is associated with attention decline in children along with severe complications of cognitive functions such as impaired visual perception and memory [73]. A study done by National Health and Nutrition Examination Survey (NHANES) (1999-2002) reported that lead blood levels is associated with increased risk of attention deficient hyperactivity syndrome in children [74]. In India, similar study was performed that reported the detrimental effects of lead on neurobehavioral functions including attention and cognitive domain [75]. Investigation done by Dietrich *et al.* also highlighted the toxic effects of lead exposure during the prenatal and postnatal period on motor development with pronounced effect on gross motor functioning [76].

Occupational exposure to lead is associated with poor visuomotor coordination, impaired visual and verbal memory performance and difficulties in decision making in adults [27]. In adults, chronic lead exposure is also linked to multiple organ dysfunctions including cardiovascular disorders and renal dysfunctions. Lead poisoning is also linked to decreased fertility, musculoskeletal defects, cataracts and brain disorders. Studies have reported that prolonged lead exposure produces disrupt muscular coordination, convulsions and coma. Lead target several important enzymatic mechanism in living body including those involve in heme synthesis that later produce severe complication [77]. Agency for Toxic Substances and Disease Registry (ATSDR) suggest that the effects of chronic exposure of Lead may persist long even after exposure ended [78]. A magnetic resonance imaging (MRI) study of former lead workers reported that high lead levels were associated with decreased total brain volume, reduced white matter volume in parietal lobe and lower volume of gray matter in the cingulum and insula [79].

CADMIUM

Mechanism of Cadmium-Induced Neurotoxicity

Cadmium is a heavy metal and is employed primarily in electroplating industry, in rods of nuclear reactors to modulate atomic fission, in manufacturing of batteries, alloys and plastics [80]. The cadmium lacks biodegradability and has an extended half-life. An estimated half-life of cadmium is almost 30 years in humans [81]. Occupational exposure generally involves employees of battery factories, electroplating industry and pigment factories [82]. Environmental exposure involves various processes in nature through which general population is exposed to cadmium such as consumption of contaminated food, leaching from cadmium containing plumbing structure, and from poorly made toys and jewelry [83]. A major route of exposure is tobacco smoking as the plant hoards cadmium and its levels in leaves are considerably increased so once tobacco is smoked; cadmium accumulates in lungs leading to deposition in blood and kidney [84].

The exact mechanism of cadmium-induced neurotoxicity is unclear, however, a few suggested pathways have suggested that the action of cadmium on central nervous system starts from blood brain barrier. Cadmium crosses adult blood brain barrier with relative difficulty as compared to younger population where it can easily crosses the barrier probably due to lack of fully functional barrier. Its presence over a longer period of time can further alter the integrity of blood brain barrier, weakening the first line of defense of brain [85]. On cellular level Cd can cause up regulation and expression of heat shock protein (*hsp60*) [86]. It interferes with uptake/transfer of calcium across cellular membrane through voltage-gated Ca++ channels leading to irregular distribution of Ca++, this effect leads to plethora of disturbances ranging from alteration in neurotransmitter discharge to induction of cell death [87, 88]. Once enters inside of the cell, cadmium interacts with mitochondrial regions leading to collapse of mitochondrial membrane potential and disrupt ATP synthesis due to interference in electron transport chain. It further causes disruption of mitochondrial membrane by mutation in mitochondrial DNA. This in turn leads to functional deficits and production of reactive species, which then additionally affect membrane fluidity of lipid bilayer through the process of lipid peroxidation which further upsets Ca++ balance across cell continuing the vicious cycle that will ultimately lead to cellular death [86, 89].

Cadmium-Induced Neurological Disorders: Clinical Evidences

Being a heavy metal, cadmium possesses extended half-life in living system. It accumulates in a biological system which then leads to various disorders. Therefore, an apparent association exists between cadmium exposure and various neurological, behavioral and psychiatric disorders [88]. Viaene and co-worker conducted a study in 2000 based upon survey and tests of a population consisting of people with chronic occupational cadmium exposure. The results revealed that the cadmium exposure is associated with reduced psychomotor activities, and impairment in attention and steadiness [90]. These results are also reflected in a study carried out in 1989 by Hart and colleagues where population under study consisted of employees working in a refrigerator coil manufacturing factory. The results showed that cadmium exposure is directly linked with reduced scores on behavioral analysis such as memory, psychomotor vigilance, and showed deficits in attention and impaired learning of new tasks [91]. As cadmium is suggested to be a more potent neurotoxin of younger population, therefore, researchers around the globe are working to determine the hazard it can impose on children. A cohort study carried out in Hubei province of China screened out 109 pregnant women for blood cadmium levels in umbilical, placental blood, and whole blood samples. The IQ of the offspring was assessed at age of 4.5 years. The results showed that cadmium levels in maternal blood during pregnancy were inversely linked to IQ of children suggesting a possible effect of cadmium on neonatal brain development [92]. Further validation of these results can be seen through another study carried out in Bangladeshi population where maternal blood cadmium levels during pregnancy were again conversely linked to intelligence scores of children later in life [93]. Moreover, a cohort study was conducted in Korea in which cadmium levels of pregnant women were determined and the IQ level of newborns was assessed at the age of 60 months, showing that increased cadmium levels in maternal blood were directly associated with lower IQ of children later in life [94]. The profound neuropsychological effects of cadmium can also be seen through a study consisting of children between the ages of 5-16 years showed a correlation between cadmium levels and cognitive deficits [95]. Such alterations were also seen in a case report of two train conductors who fought against a fire ignited in a battery box which was made up of nickel and cadmium electrode. Both of the conductors faced extensive fumes. When they were later assessed for neurological alterations then neuropsychological, memory and cognitive deficits were seen whereas urinary cadmium and nickel levels were also increased [96].

Cadmium accumulation and its effect on cognition and memory suggest a link between cadmium exposure and progression of Alzheimer's disease (AD). Since cadmium is accumulated in living system chronically and many neurological diseases like Alzheimer's disease can develop over the course of an extended period of time. Blood cadmium records and other health conditions of 51 adults who died with Alzheimer's disease were studied in a cohort study. The researchers found a link between increased blood cadmium levels and severity of disease in study subjects [97]. Parkinson's disease also poses a serious threat to public health especially in older population. A case reported in 1997 showed a potential link between cadmium and Parkinson's disease when a 64 year old man inhaled fumes of cadmium after which blood cadmium levels increased leading to metal fume fever which later caused development of lesion in brain and development of Parkinson's disease which was refractory to anti-tremor drugs [98]. Cadmium also shows association with Amyotrophic lateral sclerosis. An employee of a battery factory died from Amyotrophic lateral sclerosis after working for nine years in a cadmium exposure environment. Electromyographic studies showed injury in spinal cord and clinical progression of the disease until the man finally died at the age of 44 years [99]. Although various case reports suggest the cadmium as a potential cause of neurological diseases, however, further studies defining a link of cadmium and neurological disorders are required to pave way for new therapeutic strategies for managing these disorders.

NEUROLOGICAL DISORDERS INDUCED BY ESSENTIAL METALS IRON

Physiological Role of Iron

At physiological level, iron is involved in various neurological processes such as DNA replication, mitochondrial respiration, oxygen transportation, neurotransmitter synthesis, and myelin production [100]. For maintenance of these actions iron is required in an adequate amount. However, misregulation of iron homeostasis may lead to neurotoxicity through different mechanisms. Iron circulates in blood in the bound form with transferrin and enters into the brain cells through transferrin receptor 1 with the help of endocytosis. It releases into cytoplasm from endosomes and incorporated with ferritin, which is iron containing protein. Ferritin is soluble, non-toxic, bioavailable form and involve provide iron readily for various physiological function [101]. The neurotoxicity is induced in the condition when the concentration of iron exceeds the capacity of storage protein and results in oxidative damage and cell death. The iron homeostasis is under stringent control, however, iron gradually accumulates in the brain with advancing age as a result a portion of iron remains redox-active. The reactive species derived from iron accumulation may result in pro-toxin bioactivation, abnormal cell signaling, dysfunction of bioenergetic mechanism, proteosomal abnormality, protein aggregation, and electrophysiological instabilities leading to synaptolysis, apoptosis and necrosis [102].

Mechanism of Iron-Induced Neurotoxicity

The accumulated free iron can convert into hemosiderin and oxyhydroxide compounds. These compounds are powerful enough to exchange the electrons from the surroundings biomolecules resulting in the propagation of free radicals, leading to lipid peroxidation. The iron $Fe+2$ ion undergoes Fenton reactions resulting in the formation of strong reactive and unstable species of hydroxyl free radicals ($OH^.$). To attain the stability, these species then gain the electrons by attacking the double bond of polyunsaturated fats of lipid membrane. This results in the formation of organic free radicals which then react with molecular oxygen to produce another strong reactive specie of oxygen radical. These radicals continue the production of organic free radicals, resulting in self-destruction of lipid membrane and disruption of structural integrity of cell. This destructive pathway is continuously disseminated due to excessive free reactive iron and can potentially lead to neuronal apoptosis (Fig. **4**). Thus the accumulation of iron results in brain tissue damage making it more vulnerable to pathological conditions [103].

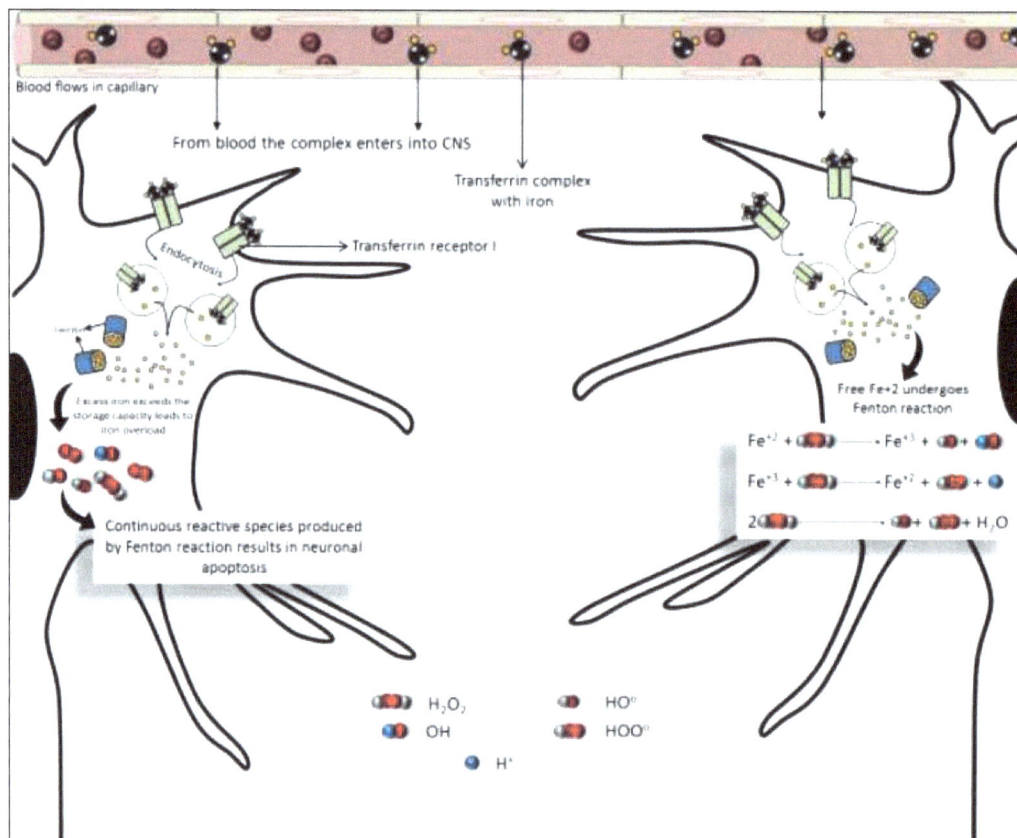

Fig. (4). Mechanism of action of iron-induced neurotoxicity. Iron flows in blood capillaries in the form of complex with transferrin protein. This complex gets access to the brain cells through transferrin receptor where iron is released from transferrin and binds with ferritin protein (storage form of iron in cell). In iron toxicity, the excess iron exceeds the capacity of storage leading to accumulation of iron and continuous production of reactive species through Fenton reaction which ultimately leads to neuronal apoptosis.

Iron-induced Neurological Disorders: Clinical Evidences

Basic MRI studies have been conducted to demonstrate the frequency of iron deposition in various brain regions in humans at different ages. Hallgren and Sourander [104] observed brains of 81 normal individuals during postmortem. They demonstrated that iron accumulation gradually increase in brain until third decade of life and attain a plateau until sixth decade of life and then again start increasing slowly in whole brain. Later on, Thomas *et al* [105] showed that considerably more iron deposition occurs in substantia nigra than other parts of the brain during first three decades of life. Moreover, iron deposition has also been observed in globus pallidus in 6 months old infants. Substantia nigra and red nucleus showed accumulation of iron at the age of 9-12 mo and 3-7 mo, respectively. Moreover, accumulation of iron in dentate gyrus has been demonstrated to occur at the age of 3-7 years of age [106]. The difference in the intensity of iron accumulation among different gender of the same age has also been demonstrated. Bartzokis (2007) [107] showed relatively lower iron disposition in women than men in different brain areas. This can explain the reason of occurrence of Alzheimer's disease and Parkinson's disease in women at a more advanced age as compared to men [107, 108].

Accumulation of iron and resultant oxidative stress has been reported to induce a list of acquired human neurological disorders including Alzheimer's disease, Parkinson's disease, amyotrophic lateral sclerosis, and multiple sclerosis [109]. It is hypothesized that increased iron content is epiphenomenon that contributes in various neurological disorders. MRI studies have strongly related brain iron deposition with age-related motor and cognitive dysfunctions [110]. Increase in iron deposition is suggested to increase the risk or worsen the developing condition of Alzheimer's disease. House *et al* has shown increased iron content in temporal gray matter in individuals suffered from memory dysfunction than normal controls. Elevated levels of iron in caudate and putamen have also been demonstrated by Bartzokis *et al* [107] in Alzheimer's patients as compared to healthy individuals. Moreover, postmortem studies of brain tissue from Alzheimer's patients showed occurrence of iron deposition. Collectively these data demonstrated direct correlation of iron accumulation with pathological condition of Alzheimer's disease. The iron chelating studies also validate these human pathological studies. Intramuscular administration of desferrioxamine, an iron chelator, to Alzheimer's patients for 24 months showed improved daily living skills as compared non-treated and placebo treated control patients. Clioquinol, another chelator of iron, has shown inhibition of iron binding to beta amyloid in rat model of Alzheimer's disease. Based on these findings clioquinol was tested in 18 Alzheimer's patients for 36 weeks by Ritchie *et al* [111] and they observed significantly low levels of beta amyloid in plasma and relatively better cognitive ability as compared to placebo treated Alzheimer's patients. These data provide a link for therapeutic targets for the treatment of neuropathological condition like Alzheimer's disease.

Iron accumulation has also been demonstrated in substantia nigra of patients who suffered from Parkinson's disease. The use of MRI, transcranial ultrasound, and histochemical methods in postmortem and *in vivo* studies has established the role of iron deposition in Parkinson's patients. Berg and Hochstrasser [112] revealed 25-100% increase in substantia nigral-iron content in Parkinson's disease when compared with healthy controls. Moreover, up-regulation of iron-associated proteins and receptors has been shown in striatum and substantia nigra of Parkinson's patients [113, 114]. Bartzokis and colleagues [115] observed difference in age for iron disposition and the onset of Parkinson's disease. They showed direct relation of increased iron content in basal ganglia and young-onset of Parkinson's disease as compared to normal controls. The accumulation of iron is suggested to induce increased susceptibility of dopaminergic neuron in substantia nigra due to free radical-induced oxidative stress and thereby increase the chance of disease occurrence [113, 114, 116]. There is varying observation regarding the iron content in Lewy body in substantia nigra. Dexter *et al* [117] did not observe increased iron levels in Lewy bodies. However, Zecca and co-workers [118] have shown the occurrence of increased iron levels in Lewy bodies using transcranial sonography, they suggested that iron may be involved in the initiation and propagation of disease condition. Trials for iron chelation therapy validate these epidemiological studies and suggested that the treatment with iron chelating drugs can improve the motor coordination in Parkinson's patients [119, 120].

The impact of iron deposition on the pathophysiology of multiple sclerosis has been frequently studied in human subjects. The association of iron deposition with the severity of multiple sclerosis has been observed from postmortem histology and *in vivo* MRI examination of human brain. The MRI changes observed in multiple sclerosis patients suggested the accumulation of iron in gray matter areas including the thalamus, red nucleus, lentiform nucleus, dentate nucleus, rolandic cortex, and caudate when compared with healthy controls of same age group [121 - 125]. The postmortem examination of brain of multiple sclerosis patients also revealed the occurrence of iron deposits in various brain regions such as putamen and thalamus [126] as well as in different cell types including neurons, oligodentrocytes, macrophages, and microglia [127, 128]. Some pre-clinical studies have demonstrated the usefulness of iron chelation therapy to reduce the severity of multiple sclerosis during active disease [129, 130]. However, studies on human subjects are warranted to validate the effectiveness of iron chelation in the disease course of multiple sclerosis.

CHROMIUM

Physiological Role of Chromium

The essential role of chromium in living body was deciphered in last decades of 21st century [131]. Chromium exists in three ionic forms in environment including 0, +3 and +6. The physiological role is associated with +3 state of chromium. The role of chromium in the regulation of lipid, carbohydrate and protein metabolism has been demonstrated previously [132]. The supplementation of chromium in various animal models showed decreased total cholesterol, low-density lipoprotein (LDL)-cholesterol, triglycerides, and increased the levels of high-density lipoprotein (HDL)-cholesterol [133]. This physiological action may be carried out by decreasing the level of leptin hormone and thus chromium is also involved in regulation of food intake. Chromium also increased the cellular uptake of amino acids. Its involvement in insulin action and glucose homeostasis is also well established [134]. Lower chromium (III) levels have been observed in obese and diabetic patients as compared to healthy controls [135]. The high concentration of chromium exists in kidneys, spleen, and liver whereas relatively lower levels are found in pancreas, muscle, heart, bones, lungs, and brain. The physiological role of chromium in central nervous system has not been completely established yet. However, the involvement of chromium in insulin action can suggest that it could impact the brain function. The physiological functions are mostly associated with +3 state of chromium, whereas, the environmental exposure to chromium (VI) may lead to toxicological and deleterious effects in a living system [136]. The expel of chromium (VI) into the environment is due to the contamination of water from industrial processes and waste, including electroplating chromium smelting, leather tanning, and paint pigments. Abnormally high concentration of chromium has been reported in the blood, body tissues and urine of workers working in welding industries.

Mechanism of Chromium-Induced Neurotoxicity

It has been suggested that hexavalent chromium is not cytotoxic itself but the oxidative reaction initiated by this transition metal produces cell damaging products [137]. The chromium (VI) readily enters into many types of cells under physiological conditions and can be reduced to trivalent state using hydrogen peroxide and antioxidant enzyme system involving reduced glutathione, glutathione reductase, and ascorbic acid [138]. During this conversion potent free radicals including hydroxyl radicals are synthesized using Fenton/Haber-Weiss mechanism which can induce damage to protein, DNA, and membrane lipid and cause cellular deterioration (Fig. **5**). The DNA damage due to hydroxyl radicals is exacerbated by altering DNA bases, depurination, DNA breaks and abnormal DNA-protein inter-linkages.

$$H_2O_2 + O_2^- \xrightarrow{Cr^{6+}} O_2 + OH^- + HO^\circ$$

The mechanism of neurotoxicity has been investigated in mice model and it has been found that exposure to chromium induces significant production of mitochondrial derived free radicals. These radicals are then scavenged by endogenous antioxidants including reduced glutathione, ascorbic acid, and α-tocopherol. The membrane polyunsaturated fatty acids are converted into free fatty acids leading to increased lipid peroxidation and oxidative burden. The antioxidant enzyme system become activated to scavenge free radicals and reduce the oxidative stress.

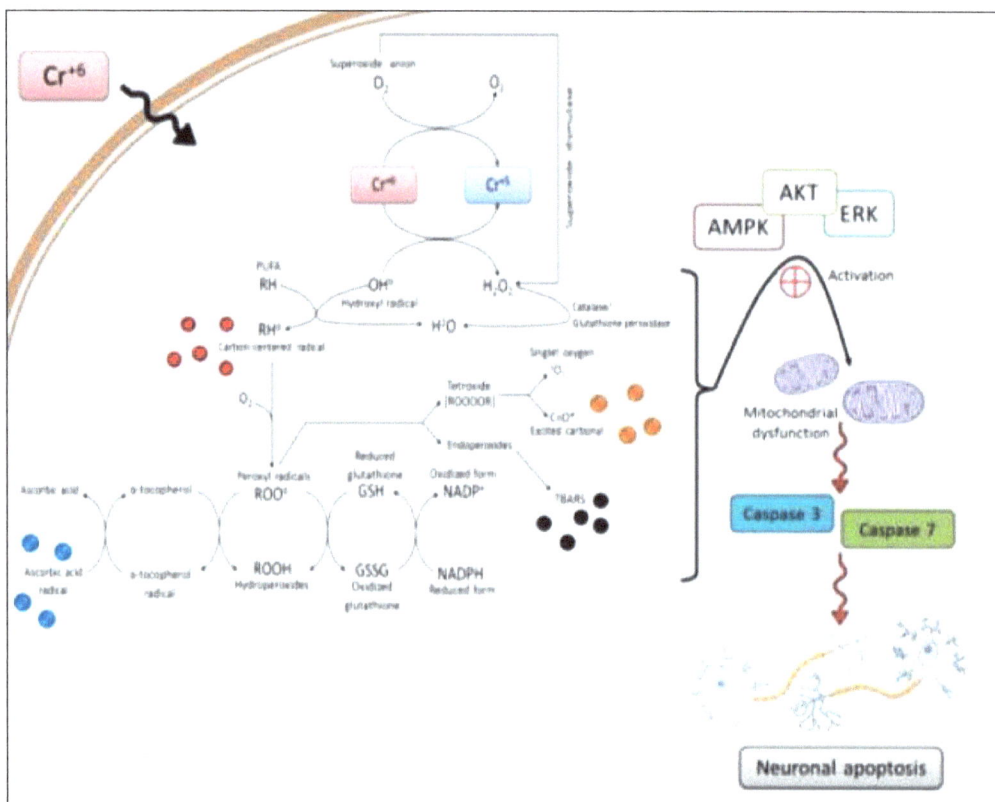

Fig. (5). Proposed mechanism of chromium-generated brain oxidative stress, antioxidant defenses against reactive oxygen species, and lipid peroxidation [138]. Refer to the text for detail.

induced by chromium exposure [138]. Recently Fu *et al* have investigated the detailed mechanism of chromium (VI) neurotoxicity. They have reported the activation of Akt/ERK/AMPK signaling pathway induced by chromium exposure in neuronal cells. This activated pathway then results in stimulation of Caspase-3/-7 pathway leading to neuronal cell death [139]. Activated AMPK pathway has been demonstrated in clinical patients of developing Alzheimer's disease [140]. The Huntington's disease and Parkinson's disease have also been associated with hyperactivated AMPK pathway [141, 142].

Chromium-Induced Neurological Disorders: Clinical Evidences

The chromium (VI) toxicity has become a serious concern for human health in most of the industrial countries. Besides affecting the respiratory, reproductive, gastrointestinal, hepatic systems, the toxicity of chromium can also disturb brain functions. Earlier clinical reports have shown the symptoms of dizziness, headache, and weakness in workers working in chrome plating plant [143]. The studies have demonstrated the imbalance of physiological chromium concentration in autistic condition. Wecker *et al* [144] and Adams *et al* [145] reported decreased chromium levels in hairs of children suffering with autism. Yorbik and co-workers [146] showed two fold increased chromium concentration in urine of autistic children. They suggested that an unusual metabolic pathway of chromium in these children and linked it with adverse oxidative stress due to decreased concentration of endogenous antioxidants including reduced glutathione, ascorbate, and lipoic acid [147 - 150].

Case study of three patients who suffered with encephalopathies also showed association with chromium exposure. The source of exposure was found to be radiological contrast agents administered to these patients, potassium chloride replacement solution and administration of mylanta, an antacid [151]. Somatopsychological effects including increased vivid, frightening dream activity, altered sleep pattern, insomnia, and irritability have also been reported in patients supplemented with chromium [152, 153]. The exposure to chromium can also occur through medical procedures such as the use of metal-on-metal (MoM) hip prostheses during joint replacement surgery. Chromium along with cobalt are the common components use in theses medical devices. It has been observed that chromium and cobalt are released by joint prostheses and enter into the blood stream [154]. Chromium can cross the blood brain barrier and deposit in the brain. The use of MoM prostheses has been linked to metal toxicity and neurological deficits may be by inducing neuronal cell death by hypoxia inducible [155 - 157]. Clark and coworkers [154] found 5-10 times higher blood chromium concentration (1.42 µg/L) in the patients who were implanted with joint prostheses and they also found subtle structural changes in the visual pathways and basal ganglia in these patients during MRI scanning of brain. Though, exposure to moderately low level of chromium for long duration did not induce defect in visual and auditory functions [158]. However, these researchers suggested that failure of MoM implants may result in very high levels of metals leading to neurological deficits as reported in other case studies [159, 160]. This was further showed in another clinical study conducted on patients implanted with MoM joints reported neurocognitive and depressive symptoms. These patients faced failure of MoM implant and were going through revised surgery. The researchers found direct correlation of neuropsychiatric symptoms with increased blood concentration of chromium (>30 µg/L) as well as cobalt [161].

Children are suggested to be more vulnerable to chromium exposure in for the neuropsychological development. Study conducted in Spain by Caparros-Gonzalez and colleagues in children living in a coastal industrial region. They found 10 fold increase in urine chromium concentration which was directly associated with neuropsychological deficits. These children showed lower IQ level, impaired selective and sustained attention, and exhibited symptoms of attention-deficit/hyperactivity disorder. Moreover, these findings were gender specific revealing more susceptibility of boys towards neuropsychological disturbances due to chromium exposure. The possible mechanism for these outcomes might be due to cell apoptosis, oxidative stress, interruption in cellular signaling pathways and DNA structural damages induced by chromium exposure [162].

COBALT

Physiological Role of Cobalt

Cobalt is another essential trace element which is required in biological system for important enzymatic reactions. The vitamin-B12 is the active form of cobalt in which cobalt makes a complex with four pyrrole ring structures. Vitamin-B12 or cynocobalamin acts as a coenzymes in various enzymatic processes such as formation of methionine, metabolism of purines, and of methylmalonic acid in succinic acid [163]. Cobalt is used in the production of color pigments, steel and alloys [164]. It is also used as catalyst in chemical and oil-refining industries, electroplating, and as additive in fertilizer and animal feed. In medical field cobalt is used in radiology and artificial joints implants [165]. Diet is the main of source of cobalt exposure in general population whereas inhalation represents primary source of occupational exposure [166].

Mechanism of Cobalt-Induced Neurotoxicity

Researchers have investigated the mechanism of cobalt-induced neurotoxicity in cultured neurons (Fig. **6**). The exposure of cobalt in the growing neuronal cell mediated the production of reactive species by reacting with H_2O_2 through Fenton-reaction. These cobalt-mediated reactive species can hit various cellular pathways, such as mitogen-activated protein kinase (MAPK) pathway, mTOR, and inflammatory responses. Lan and colleagues observed the production of reactive radicals and activation of MAPK following the treatment of neuronal cells with $CoCl_2$ [167]. The activated MAPK then activates the downstream signaling molecules leading to the stimulation of caspase 3 and apoptotic events [168]. Repression of mTOR signaling has also been reported in cultured neuronal cells after the treatment of $CoCl_2$ [169]. Inflammatory mediators such as interlukin-6 are also found to be up-regulated in $CoCl_2$-treated neurons leading to increased inflammatory responses and ultimately cell death [170].

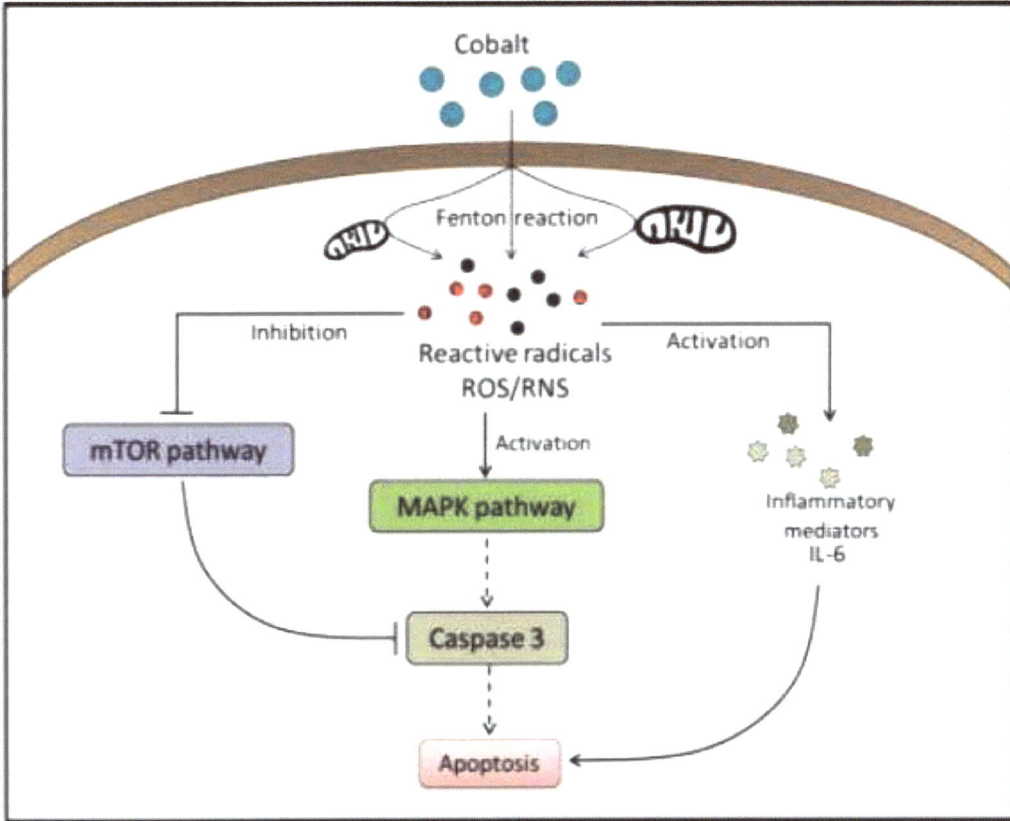

Fig. (6). Postulated mechanism of action of cobalt-induced neurotoxicity.

Cobalt-Induced Neurological Disorders: Clinical Evidences

Occupational exposure provides the main source of cobalt exposure in human population. The workers exposed to cobalt have shown neurological deficits such as memory loss, neuropathies and visual defects. Jordan and workfellows reported impaired attention and verbal memory loss in workers working with cobalt, this was accompanied with loss of visual-spatial memory [171]. Long term exposure to cobalt has been associated with visual and auditory deficits. A-middle aged worker exposed to cobalt dust for 20 months developed bilateral optic atrophy and bilateral deafness. Interestingly, these symptoms started dampening after 1 month, and became completely normalized after 14 months of discontinuation of work in cobalt dust environment [172]. Cobalt toxicity has been reported due to its medicinal use for the treatment of anemia, nephritis and infections [173]. Gardner reported tinnitus in 5 patients administered with $CoCl_2$ for 4-16 weeks for the treatment of anemia and uremia. One out of these patients developed complete deafness after 12 weeks of treatment. These symptoms co-occurred with paraesthesia in the limbs, absent tendon jerks and calf tenderness indicated that the patient had a peripheral neuritis and suffered with difficulty in walking [174]. Another case report of a patient treated with $CoCl_2$ showed similar symptoms of paraesthesia, impaired walking pattern, and loss of hearing representing bilateral nerve deafness, and loss ankle reflexes. All these symptoms were recovered within 4 months of termination of cobalt treatment [175]. Visual changes following cobalt therapy have also been reported in other case studies which were reversed upon cessation of treatment indicating that cobalt toxicity exerted reversible disturbances in the nervous system [176].

There are considerable case studies reporting neurological symptoms following the use of joint implants having cobalt as the main component. Megaterio *et al* [177] observed sensory-motor polyneuropathy in patient after one of implantation of MoM prostheses. Optic atrophy and retinopathy, bilateral deafness was also observed after two years of joints implants [178]. Oldenburg and coworkers reported neuropathy, convulsions and headache in a patient surgically implanted with metal prostheses [179]. Cranial nerve impairment and mild distal sensory-motor disturbances have also been reported in such patients [159]. Moreover, hearing loss, tinnitus, tremors, poor motor co-ordination, and impaired memory have been observed in patients implanted with MoM prostheses [160, 180]. All these clinical findings reported several folds of elevated cobalt levels in serum, plasma and blood samples of patients [181].

COPPER

Physiological Role of Copper

Copper is the third essential trace metal after zinc and is found in the oxidized Cu (II) and reduced Cu (I) states in all living organisms [182]. The WHO experts panel committee has recommended a daily dose of 2mg/kg copper for an adult [183]. In mammalian tissues, it is an obligatory component of eight enzymes, required for lysyl oxidase in connective tissue, dopamine monooxygenase in brain, ceruloplasmin, and cytochrome c oxidase, the last enzyme is in the electron transport chain in the mitochondria. As a cofactor it protects protein, membrane lipids, and nucleic acid from damage [184, 185].

Mechanism of Copper-Induced Neurotoxicity

Copper is required for essential body function however, it can also accelerate toxic oxidative stress [186]. Like iron, copper can take part in reactions that leads to the production of reactive species which results in lipid membrane peroxidation, direct protein oxidation, and nucleic acid molecules cleavage [187]. The production and reaction of reactive species with different biomolecules in human body contribute towards the development of different diseases of nervous system, aging, and cancer [188]. In brain copper is prominent in hippocampus, basal ganglia, cerebellum, cerebellar granular, and cell bodies of cortical pyramidal neurons and numerous synaptic membranes [189]. Disturbance of copper homoeostasis in brain may cause several diseases. For instant, excess amount of copper in brain causes accumulation of this metal in different organs like liver, basal ganglia (Wilson's disease), and results in organ dysfunction which may lead to death [190]. Similarly, deficiency of brain copper is connected to Menkes (a genetic disorder caused by the mutation of ATP7A copper transporter gene) and Alzheimer's disease [191 - 193].

Copper-Induced Neurological Disorders: Clinical Evidences

Copper is required for various physiological functions and it is present in high concentration in brain. Since excess copper can easily accumulate at the natural site of storage, therefore, the essentiality of copper makes this metal as one of the causing factors to induce neurological disorders. Experimental findings have reported impaired copper transportation and abnormal copper-protein interactions in human brain diseases providing the evidence of critical importance of copper in normal neurological functions [194]. The neurological disorders, in which copper has been implicated directly or indirectly, include Alzheimer's disease, Parkinson's disease, autism, Huntington disease, and prion disease [195]. A reduction in copper concentration has been linked with Alzheimer's disease however, there are clinical studies reporting excess amount of copper in such patients [196]. The increased aggregation of β-amyloid protein, a pathogenic protein present in Alzheimer's brain, has been observed due to binding of copper with this protein [197, 198]. Miller and colleagues analyzed the concentration of metal ions (including copper) in amyloid plaques of Alzheimer's brain tissues. They found significantly increased accumulation of copper in plaques using synchrotron X-ray fluorescence microprobe technique [199]. Before Miller's team, Lovell and coworker [200] also reported accumulation of copper in senile plaques in Alzheimer's patients. Other studies [201, 202] including a meta-analysis [203] also reported dysregulation of copper in brain of Alzheimer's patients. These authors reported increased levels of non-ceruloplasmin-bound copper concentration in serum sample of Alzheimer's patients. Furthermore, meta-analyses published in last ten years strongly associated elevated total- and free-copper levels in serum-plasma samples of Alzheimer's patients [204 - 208]. These results were are observed in other studies reporting increased copper (Cu^{++}) concentrations in Alzheimer's patients as compared to healthy controls [209 - 211].

Occupational exposure of copper for a long duration has also been linked with the occurrence of Parkinson's disease. A case-control study was performed by Gorell and colleagues on 144 Parkinsonian patients in Detroit city of US who were exposed to heavy metals for more 20 years. It was found that long term exposure to copper alone or in combination with lead or iron may result in Parkinson's disease [212]. Prevalence of Parkinson's disease in urban population was found to be directly associated with industrial emission of copper in to the environment [213]. These results can be further linked with the clinical studies reporting increased concentration of copper in biological samples of Parkinson's patients. Elevated copper levels in serum [214], cerebrospinal fluid [215, 216] and even in brain [217] have been reported previously. Thus, copper is considered as an important risk factor for the disease onset [170]. Copper toxicity has also been associated with autism, a neurodevelopmental disorder characterized by social deficits, language impairment, and repetitive behaviors. The autistic symptoms appear in the first three years of life. Since the last two decades there is an increased occurrence of autistic cases [218, 219]. The exact cause and etiology of the disease has not been identified yet, however, researchers have postulated a potential association between copper levels and autism. Russo [220] reported significantly elevated levels of plasma copper in 79 autistic adult patients as compared to healthy individuals. Elevated expression of metallothioneins and increased oxidative stress have also been found in the blood of twenty eight children suffering from autism spectrum disorder, representing a critical role of heavy metals in the development of autistic symptoms [221].

Copper accumulation has been observed in the human brain of Huntington's disease. Dexter and coworker analyzed the trace metal content in autopsy studies of brain samples of patients died with Huntington's disease. They found significantly increased total copper content in the putamen region of Huntington's disease brain [222]. Later on, Fox and colleagues conducted an experiment on brain samples obtained from postmortem of Huntington's patients and found significantly increased copper levels. They also performed some pre-clinical studies in Huntington's mice model and reported impaired redox balance and increased interaction of copper with Huntingtin protein, a pathogenic protein accumulated in Huntington's brain. These authors postulated that copper can promote aggregation of Huntingtin protein leading to disease progression as observed in the case of β-amyloid protein [223]. Mutation in copper transporter (Ctr-1) has also been reported previously, further emphasizes a critical of copper in Huntington's disease [224]. Copper is also found to have affinity with other cellular protein such as prion protein. In pathological condition abnormal prion forms protein aggregates leading to a disease state of prion disease. Brown [225] reported increased levels of copper in human subjects of prion disease. Pre-clinical data also showed that copper can bind to prion protein with high affinity which may play an important role in disease progression [226].

MANGANESE

Physiological Role of Manganese

Manganese is the trace element which acts as a co-factor in the active site of various enzymes [227]. Naturally, it is present in plant-based food including pecans, grapes, nuts, tea leaves, grains, green leafy vegetables, meats, and poultry [228]. Manganese is required for the synthesis of amino acids, protein, lipids, carbohydrates, for normal development, maintenance of nerve, the function of immune cells, and for vitamins and blood sugar regulation [229, 230]. Inadequate manganese consumption in animal and human models showed reproductive defects, loss of weight, impaired growth, dermal issues, and altered mood [231, 232]. Overexposure to manganese may result in the accumulation of this trace element in the basal ganglia and therefore, can cause adverse neurotoxic effects [233]. Three routes of manganese absorbance in human body have been reported *i.e*, through the gastrointestinal tract, lungs following inhalation exposure, intravenous injection *via* illegal manganese narcotics [234]. After absorption it enters into the blood stream and then from olfactory tract it enters into the brain [235, 236].

Mechanism of Manganese-Induced Neurotoxicity

Out of eleven oxidative states, Mn (II) and Mn (III) are most biological states of manganese. Mn (III) is taken up by the cells more efficiently and accumulates more as compared to Mn (II). The cellular effects of Mn (III) are more toxic than Mn (II) [68]. In brain, the excess amount of manganese mainly accumulates in the substantia nigra, globus pallidus, caudate-putamen, and striatum [227, 237]. Inside the neurons, it enters into the mitochondria and hinders the synthesis of ATP by electron transport system [238]. This happens due to the inhibition of ATP synthase or complex I (NADH dehydrogenase) [239, 240]. Inhibition of complex II (succinate dehydrogenase) or glutamate/aspartate exchanger is also reported due to manganese toxicity [241]. The disruption of ATP synthesis may lead to accumulation of reactive species and ultimately oxidative stress [242]. Previous experimental studies particularly reported the disruption of dopaminergic system due to manganese toxicity which may result in idiopathic Parkinsonism [243, 244] (Burton and Guilarte, 2009; Burton *et al.*, 2009). Manganese converts dopamine into reactive quinone species which further induce additive effects in manganese-induced oxidative stress [245]. Manganese induces impairment in dopamine release and reduces the function of D2 receptor [243]. It also interacts with dopamine transporter and interferes with the pre-synaptic events [246]. Gamma aminobutyric acid (GABA)-ergic and glutamatergic systems have found to be unaffected due to manganese toxicity, consequently, it has been suggested that dopaminergic degeneration may not occur in manganese toxicity but instead result in dopamine dysfunction and result in idiopathic Parkinsonism [68]. The behavioral and cognitive deficits observed in manganese toxicity, therefore, related to the interference in dopaminergic system [247] (Fig. 7). The cellular homeostatic system of manganese is highly efficient. However, due to continuous exposure to high manganese, the system check is not maintained resulting in increased uptake of this metal inside of cell. Continued uptake of manganese leads to mitochondrial dysfunction which then releases cytochrome c and activate caspase 9 and caspase 3 resulting in DNA fragmentation and apoptosis [248].

Manganese-Induced Neurological Disorders: Clinical Evidences

The neurotoxicity-induced by excessive exposure to manganese may result in clinical condition of manganism, which is characterized as extra-pyramidal neurological disorder. This clinical manifestation shows similar symptoms of Parkinson's disease, since manganese toxicity particularly affects dopaminergic system. The symptoms of manganism include muscular rigidity, tremor, bradykinesia, gait disturbances, memory dysfucntions, mood disorder and a mask-like expression. However, there are few differences between these two clinical conditions with the major emphasis is given to insensitivity of manganism patients towards L- dihydroxyphenylalanine (L-DOPA) treatment [249 - 252]. This is because of the fact that manganese toxicity affects dopamine release and transport, the condition in which L-DOPA treatment becomes ineffective.

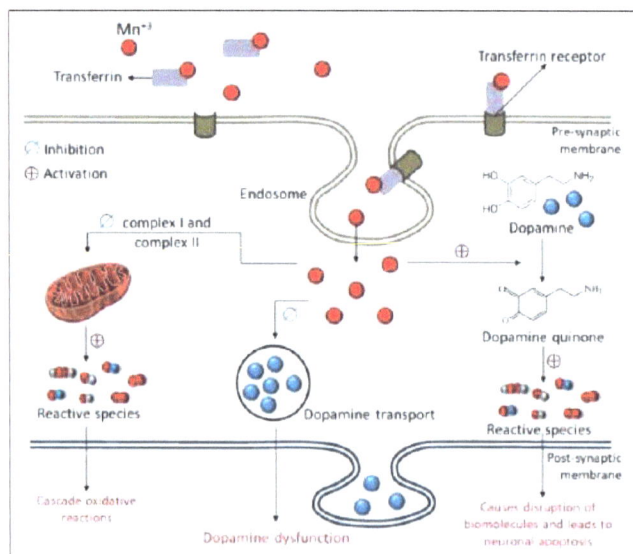

Fig. (7). Mechanism of action of manganese-induced neurotoxicity.

Brain imaging techniques are used to analyze deposition of manganese in humans who are occupationally exposed to manganese. The accumulation of manganese in basal ganglia has been found in workers occupationally exposed to airborne manganese using T1-weighed magnetic resonance imagining technique [253, 254]. Using single-photon emission computed tomography (SPECT) and positron emission tomography (PET), it has been shown that elevated brain manganese can result in deficits in the dopaminergic system in exposed humans [253]. However, a careful interpretation has been suggested for such cases as underlying Parkinson's disease can influence the results [68]. Dysfunction of dopaminergic system following manganese toxicity has also been found in addicted youngsters who were intoxicated with very high levels of manganese by injecting homemade psychostimulant preparations (ephedron). These young drug addicts exhibited clinical symptoms of Parkinsonism, showed insensitivity to L-DOPA treatment and normal levels of dopamine transporter levels in the striatum as observed in SPECT studies [255 - 264]. Moreover, abnormalities in the premotor cortex and medial frontal cortex have also been found in ephedron addicts. These brain regions are involved in motor and executive functions [263]. Besides Parkinsonism, manganese toxicity has also been linked with other neurodegenerative diseases such as prion disease. The pathogenic prion protein has found to have high affinity for manganese binding. The long-term exposure to manganese induces instability in prion protein by binding with it, resulting in misfolding and prion disease pathogenesis [248]. Wong *et al.* [265] reported increased plasma and brain manganese levels in human patients suffering from Creutzfeldt–Jakob disease.

Manganese toxicity also found to induce neurotoxic effects in children. Studies have reported memory and behavioral deficits in children who were drinking water with high levels of manganese. These children also showed hyperactive behavior, aggressive behavior, and deficits in attention in school [266, 267]. Bouchard and coworkers showed reduced IQ levels in school children who were exposed to manganese by drinking manganese-rich groundwater [267]. Environmental exposure to manganese in children living in close proximity to manganese alloy plant also resulted in reduced full scale and verbal IQ scores. These results were directly related to increased levels of hair manganese in exposed children [268]. Moreover, exposure to low or high levels of manganese in early life can induce long-term neurological deficits and can negatively influence child neurodevelopment [269].

NICKEL

Physiological Role of Nickel

Nickel is an essential element for living organisms present in both soluble and insoluble form in nature with several oxidative states (from -1 to +4) [270]. However, +2 oxidation states are the most common state in the biological systems and environment [271, 272]. Most of the vegetables and fruits like green beans, peas, broccoli, dried fruit and nuts, cocoa, and chocolate contain nickel [273]. Like other metallic micronutrients, an imbalanced quantity of nickel in the human body either in deficient or excess amount leads to several physiological and behavioral changes. Its deficiency leads to growth inhibition, slow reproductive rate, glucose and lipid metabolism alteration, reduced enzymatic activity in animals, and other metal ion content alteration [274]. An excess amount of nickel at a high level also causes organ dysfunctioning and multiple toxic effects in the lungs, liver, and brain.

Mechanism of Nickel-Induced Neurotoxicity

Brain is highly susceptible to nickel toxicity and causes different pathological effects. The reactive oxygen species produced by nickel exposure play a central role in nickel-mediated neurotoxicity. Oxidative stress generated by nickel disrupts neurotransmission, reduces GABA level and alters its uptake [275]. In the nervous system, it is absorbed by the olfactory pathway or through the blood-brain barrier (BBB) failures and leads to cell toxicity in nerve cells due to the accumulation of nickel in the cerebral cortex and whole brain. Consequently, this accumulation resulted in various symptoms of the neurological disorder like giddiness, headaches, lethargy, tiredness, behavioral deficiencies, and interrupt neurotransmitter activity [276 - 278].

Nickel-induced Neurological Disorders: Clinical Evidences

Occupational exposure to nickel is fairly common in battery factories; however, the focus has mostly been given to cadmium as it is a non-essential metal which has been linked to toxicity and progression of various diseases. However, in some instances nickel has proven to be hazardous such as seen in a case of a gaseous diffusion plant where employees were exposed to nickel powder; the study displayed an increased mortality rate among workers exposed to nickel as compared to unexposed workers [279]. An association between nickel toxicity and Parkinson's disease has been found through a study conducted on samples of blood and serum taken from Parkinson's patients and healthy controls. This metalomic study displayed an increased amount of nickel in both serum and blood samples of Parkinson's patients [280]. Similarly in another study, cerebrospinal fluid samples from Parkinson's patients showed visibly increased nickel when compared with healthy controls [281]. These results are also reflected in a case-control study conducted in 2016 in which cerebrospinal fluid samples of 6 Parkinson's patients showed increased levels of nickel [282]. Thus, nickel shows a considerable link with neurodegenerative diseases specially Parkinsonism.

CONCLUDING REMARKS

Heavy metals are present ubiquitously in the environment. Though essential metals are required for the normal physiology of living beings, however, their excessive amount can lead to neurotoxic effects. Therefore, these metals should be taken in the right amount to utilize the beneficial effects and to diminish the toxicity. For hazardous heavy metals, special measures should be considered to eradicate their toxicity from the environment. Nonetheless, due to modern civilization and increased setups of industries it is very difficult to avoid metal exposure, especially in occupational workers. Therefore, the daily routine diet of general public should contain dietary elements which have chelating metal properties. More research is needed in this field to explore metal chelating dietary components to protect brain function from metal-induced toxicity.

CONSENT FOR PUBLICATION

Not Applicable.

CONFLICT OF INTEREST

The author declares no conflict of interest, financial or otherwise.

ACKNOWLEDGEMENTS

Declared none.

ABBREVIATIONS

5-HT	5-hydroxytryptamine
BBB	blood brain barrier
BDNF	brain derived neurotropic factor
CREB	cAMP response element binding protein
FDA	Food and Drug administration
GABA	gamma aminobutyric acid
HDL	high-density lipoprotein
IQ	intelligent quotient
LDL	low-density lipoprotein

L-DOPA L-dihydroxyphenylalanine

MAO-B monoamine oxidase-B

MAPK mitogen-activated protein kinase

MoM metal on metal

MRI magnetic resonance imaging

NMDA N-methyl-D-aspartate

RNS reactive nitrogen species

ROS reactive oxygen species

REFERENCES

[1]   Exley C. Aluminium and medicine.Molecular and Supramolecular Bioinorganic Chemistry: Applications in Medical Sciences. New York: Nova Biomedical Books 2009; pp. 45-68.

[2]   Gonçalves PP, Silva VS. Does neurotransmission impairment accompany aluminium neurotoxicity? J Inorg Biochem 2007; 101(9): 1291-338.
[http://dx.doi.org/10.1016/j.jinorgbio.2007.06.002] [PMID: 17675244]

[3]   Valls-Pedret C, Lamuela-Raventós RM, Medina-Remón A, *et al.* Polyphenol-rich foods in the Mediterranean diet are associated with better cognitive function in elderly subjects at high cardiovascular risk. J Alzheimers Dis 2012; 29(4): 773-82.
[http://dx.doi.org/10.3233/JAD-2012-111799] [PMID: 22349682]

[4]   Tomljenovic L, Shaw CA. Aluminium vaccine adjuvants: are they safe? Curr Med Chem 2011; 18(17): 2630-7.
[http://dx.doi.org/10.2174/092986711795933740] [PMID: 21568886]

[5]   Savory J, Bertholf RL, Wills MR. Aluminium toxicity in chronic renal insufficiency. Clin Endocrinol Metab 1985; 14(3): 681-702.
[http://dx.doi.org/10.1016/S0300-595X(85)80012-8] [PMID: 3905084]

[6]   Polizzi S, Pira E, Ferrara M, *et al.* Neurotoxic effects of aluminium among foundry workers and Alzheimer's disease. Neurotoxicology 2002; 23(6): 761-74.
[http://dx.doi.org/10.1016/S0161-813X(02)00097-9] [PMID: 12520766]

[7]   Sińczuk-Walczak H, Szymczak M, Raźniewska G, Matczak W, Szymczak W. Effects of occupational exposure to Aluminium on nervous system: clinical and electroencephalographic findings. Int J Occup Med Environ Health 2003; 16(4): 301-10.
[PMID: 14964639]

[8]   Bishop NJ, Morley R, Day JP, Lucas A. Aluminium neurotoxicity in preterm infants receiving intravenous-feeding solutions. N Engl J Med 1997; 336(22): 1557-61.
[http://dx.doi.org/10.1056/NEJM199705293362203] [PMID: 9164811]

[9]   Exley C, Esiri MM. Severe cerebral congophilic angiopathy coincident with increased brain aluminium in a resident of Camelford, Cornwall, UK. J Neurol Neurosurg Psychiatry 2006; 77(7): 877-9.
[http://dx.doi.org/10.1136/jnnp.2005.086553] [PMID: 16627535]

[10] Shoenfeld Y, Agmon-Levin N. 'ASIA' - autoimmune/inflammatory syndrome induced by adjuvants. J Autoimmun 2011; 36(1): 4-8.
[http://dx.doi.org/10.1016/j.jaut.2010.07.003] [PMID: 20708902]

[11] Aluminium in large and small volume parenterals used in total parenteral nutrition, amendment 2003.http://edocket.access.gpo.gov/cfr_2005/aprqtr/pdf/21cfr201.323.pdf

[12]   Bondy SC. The neurotoxicity of environmental Aluminium is still an issue. Neurotoxicology 2010; 31(5): 575-81.
[http://dx.doi.org/10.1016/j.neuro.2010.05.009] [PMID: 20553758]

[13]   Rondeau V, Commenges D, Jacqmin-Gadda H, Dartigues JF. Relation between Aluminium concentrations in drinking water and Alzheimer's disease: an 8-year follow-up study. Am J Epidemiol 2000; 152(1): 59-66.
[http://dx.doi.org/10.1093/aje/152.1.59] [PMID: 10901330]

[14]   Sherrard DJ, Walker JV, Boykin JL. Precipitation of dialysis dementia by deferoxamine treatment of Aluminium-related bone disease. Am J Kidney Dis 1988; 12(2): 126-30.
[http://dx.doi.org/10.1016/S0272-6386(88)80007-6] [PMID: 3400633]

[15]   Exley C, Vickers T. Elevated brain aluminium and early onset *al*zheimer's disease in an individual occupationally exposed to aluminium: a case report. J Med Case Reports 2014; 8: 41.
[http://dx.doi.org/10.1186/1752-1947-8-41] [PMID: 24513181]

[16]   Lukiw WJ, Kruck TPA, Percy ME, *et al.* Aluminium in neurological disease - a 36 year multicenter study. J Alzheimers Dis Parkinsonism 2019; 8(6): 457.
[PMID: 31179161]

[17]   Campbell A. The potential role of aluminium in Alzheimer's disease. Nephrol Dial Transplant 2002; 17 (Suppl. 2): 17-20.
[http://dx.doi.org/10.1093/ndt/17.suppl_2.17] [PMID: 11904353]

[18]   Altmann P, Cunningham J, Dhanesha U, Ballard M, Thompson J, Marsh F. Disturbance of cerebral function in people exposed to drinking water contaminated with aluminium sulphate: retrospective study of the Camelford water incident. BMJ 1999; 319(7213): 807-11.
[http://dx.doi.org/10.1136/bmj.319.7213.807] [PMID: 10496822]

[19]   Tseng HP, Wang YH, Wu MM, The HW, Chiou HY, Chen CJ. Association between chronic exposure to arsenic and slow nerve conduction velocity among adolescents in Taiwan. J Health Popul Nutr 2006; 24(2): 182-9.
[PMID: 17195559]

[20]   Prakash C, Soni M, Kumar V. Mitochondrial oxidative stress and dysfunction in arsenic neurotoxicity: A review. J Appl Toxicol 2016; 36(2): 179-88.
[http://dx.doi.org/10.1002/jat.3256] [PMID: 26510484]

[21]   Ríos R, Zarazúa S, Santoyo ME, *et al.* Decreased nitric oxide markers and morphological changes in the brain of arsenic-exposed rats. Toxicology 2009; 261(1-2): 68-75.
[http://dx.doi.org/10.1016/j.tox.2009.04.055] [PMID: 19409443]

[22]   Jing C, Cui J, Huang Y, Li A. Fabrication, characterization, and application of a composite adsorbent for simultaneous removal of arsenic and fluoride. ACS Appl Mater Interfaces 2012; 4(2): 714-20.
[http://dx.doi.org/10.1021/am2013322] [PMID: 22235839]

[23]   Kannan GM, Tripathi N, Dube SN, Gupta M, Flora SJ, Flora SJ. Toxic effects of arsenic (III) on some hematopoietic and central nervous system variables in rats and guinea pigs. J Toxicol Clin Toxicol 2001; 39(7): 675-82.
[http://dx.doi.org/10.1081/CLT-100108508] [PMID: 11778665]

[24]   Edwards M, Hall J, Gong G, O'Bryant SE. Arsenic exposure, AS3MT polymorphism, and neuropsychological functioning among rural dwelling adults and elders: a cross-sectional study. Environ Health 2014; 13(1): 15.
[http://dx.doi.org/10.1186/1476-069X-13-15] [PMID: 24621105]

[25]   States JC, Barchowsky A, Cartwright IL, Reichard JF, Futscher BW, Lantz RC. Arsenic toxicology: translating between experimental models and human pathology. Environ Health Perspect 2011; 119(10): 1356-63.
[http://dx.doi.org/10.1289/ehp.1103441] [PMID: 21684831]

[26]   Kilburn KH. Neurobehavioral impairment from long-term residential arsenic exposure.Arsenic. Dordrecht: Springer 1997; pp. 159-75.
[http://dx.doi.org/10.1007/978-94-011-5864-0_14]

[27]   Chen P, Miah MR, Aschner M. Metals and neurodegeneration. F1000Research. 2016; 5 F1000 Faculty Rev-366

[28]   Kim S, Takeuchi A, Kawasumi Y, Endo Y, Lee H, Kim Y. A Guillain-Barré syndrome-like neuropathy associated with arsenic exposure. J Occup Health 2012; 54(4): 344-7.
[http://dx.doi.org/10.1539/joh.12-0023-CS] [PMID: 22672885]

[29]   Mukherjee SC, Rahman MM, Chowdhury UK, *et al.* Neuropathy in arsenic toxicity from groundwater arsenic contamination in West Bengal, India. J Environ Sci Health Part A Tox Hazard Subst Environ Eng 2003; 38(1): 165-83.
[http://dx.doi.org/10.1081/ESE-120016887] [PMID: 12635825]

[30]   Calderón J, Navarro ME, Jimenez-Capdeville ME, *et al.* Exposure to arsenic and lead and neuropsychological development in Mexican children. Environ Res 2001; 85(2): 69-76.
[http://dx.doi.org/10.1006/enrs.2000.4106] [PMID: 11161656]

[31]   Parvez F, Wasserman GA, Factor-Litvak P, *et al.* Arsenic exposure and motor function among children in Bangladesh. Environ Health Perspect 2011; 119(11): 1665-70.
[http://dx.doi.org/10.1289/ehp.1103548] [PMID: 21742576]

[32]   Vahidnia A, van der Voet GB, de Wolff FA. Arsenic neurotoxicity--a review. Hum Exp Toxicol 2007; 26(10): 823-32.
[http://dx.doi.org/10.1177/0960327107084539] [PMID: 18025055]

[33]   Gong G, O'Bryant SE. The arsenic exposure hypothesis for Alzheimer disease. Alzheimer Dis Assoc Disord 2010; 24(4): 311-6.
[http://dx.doi.org/10.1097/WAD.0b013e3181d71bc7] [PMID: 20473132]

[34]   Berbel-García A, González-Aguirre JM, Botia-Paniagua E, *et al.* [Acute polyneuropathy and encephalopathy caused by arsenic poisoning]. Rev Neurol 2004; 38(10): 928-30.
[PMID: 15175974]

[35]   Andersen HR, Nielsen JB, Grandjean P. Toxicologic evidence of developmental neurotoxicity of environmental chemicals. Toxicology 2000; 144(1-3): 121-7.
[http://dx.doi.org/10.1016/S0300-483X(99)00198-5] [PMID: 10781879]

[36]   Gamou I. Arsenic. Intern Med 1971; 27: 582-5. [in Japanese].

[37]   Dakeishi M, Murata K, Grandjean P. Long-term consequences of arsenic poisoning during infancy due to contaminated milk powder. Environ Health 2006; 5(1): 31.
[http://dx.doi.org/10.1186/1476-069X-5-31] [PMID: 17076881]

[38]   Yamashita N, Doi M, Nishio M, Hojo H, Tanaka M. [Recent observations of Kyoto children poisoned by arsenic tainted "Morinaga Dry Milk" (author's transl)]. Jpn J Hyg 1972; 27(4): 364-99.
[http://dx.doi.org/10.1265/jjh.27.364] [PMID: 4679420]

[39]   Echeverria D, Woods JS, Heyer NJ, *et al.* The association between serotonin transporter gene promotor polymorphism (5-HTTLPR) and elemental mercury exposure on mood and behavior in humans. J Toxicol Environ Health A 2010; 73(15): 1003-20.
[http://dx.doi.org/10.1080/15287390903566591] [PMID: 20526950]

[40]   Echeverria D, Woods JS, Heyer NJ, *et al.* The association between a genetic polymorphism of coproporphyrinogen oxidase, dental mercury exposure and neurobehavioral response in humans. Neurotoxicol Teratol 2006; 28(1): 39-48.
[http://dx.doi.org/10.1016/j.ntt.2005.10.006] [PMID: 16343843]

[41]   Echeverria D, Woods JS, Heyer NJ, *et al.* Chronic low-level mercury exposure, BDNF polymorphism, and associations with cognitive and motor function. Neurotoxicol Teratol 2005; 27(6): 781-96.

[http://dx.doi.org/10.1016/j.ntt.2005.08.001] [PMID: 16301096]

[42] Heyer NJ, Echeverria D, Bittner AC Jr, Farin FM, Garabedian CC, Woods JS. Chronic low-level mercury exposure, BDNF polymorphism, and associations with self-reported symptoms and mood. Toxicol Sci 2004; 81(2): 354-63.
[http://dx.doi.org/10.1093/toxsci/kfh220] [PMID: 15254338]

[43] Heyer NJ, Echeverria D, Martin MD, Farin FM, Woods JS. Catechol O-methyltransferase (COMT) VAL158MET functional polymorphism, dental mercury exposure, and self-reported symptoms and mood. J Toxicol Environ Health A 2009; 72(9): 599-609.
[http://dx.doi.org/10.1080/15287390802706405] [PMID: 19296409]

[44] Ritchie KA, Burke FJ, Gilmour WH, *et al.* Mercury vapour levels in dental practices and body mercury levels of dentists and controls. Br Dent J 2004; 197(10): 625-32.
[http://dx.doi.org/10.1038/sj.bdj.4811831] [PMID: 15611750]

[45] Syversen T, Kaur P. The toxicology of mercury and its compounds. J Trace Elem Med Biol 2012; 26(4): 215-26.
[http://dx.doi.org/10.1016/j.jtemb.2012.02.004] [PMID: 22658719]

[46] Bjørklund G, Tinkov AA, Dadar M, *et al.* Insights into the potential role of mercury in Alzheimer's disease. J Mol Neurosci 2019; 67(4): 511-33.
[http://dx.doi.org/10.1007/s12031-019-01274-3] [PMID: 30877448]

[47] Clarkson TW. The toxicology of mercury. Crit Rev Clin Lab Sci 1997; 34(4): 369-403.
[http://dx.doi.org/10.3109/10408369708998098] [PMID: 9288445]

[48] Tchounwou PB, Ayensu WK, Ninashvili N, Sutton D. Environmental exposure to mercury and its toxicopathologic implications for public health. Environ Toxicol 2003; 18(3): 149-75.
[http://dx.doi.org/10.1002/tox.10116] [PMID: 12740802]

[49] D'ltri PA, D'ltri FM. Mercury contamination: a human tragedy. Environ Manage 1978; 2(1): 3-16.
[http://dx.doi.org/10.1007/BF01866442]

[50] Amin-Zaki L, Majeed MA, Greenwood MR, Elhassani SB, Clarkson TW, Doherty RA. Methylmercury poisoning in the Iraqi suckling infant: a longitudinal study over five years. J Appl Toxicol 1981; 1(4): 210-4.
[http://dx.doi.org/10.1002/jat.2550010405] [PMID: 6892222]

[51] Goldman LR, Shannon MW. Technical report: mercury in the environment: implications for pediatricians. Pediatrics 2001; 108(1): 197-205.
[http://dx.doi.org/10.1542/peds.108.1.197] [PMID: 11433078]

[52] Hammond TJ, Gallo CF. Effects of argon atoms on the self-absorption and the intensity of hg 2537-a radiation in hg + ar discharges. Appl Opt 1971; 10(1): 58-64.
[http://dx.doi.org/10.1364/AO.10.000058] [PMID: 20094391]

[53] Burbacher TM, Rodier PM, Weiss B. Methylmercury developmental neurotoxicity: a comparison of effects in humans and animals. Neurotoxicol Teratol 1990; 12(3): 191-202.
[http://dx.doi.org/10.1016/0892-0362(90)90091-P] [PMID: 2196419]

[54] Schwarz S, Husstedt I, Bertram HP, Kuchelmeister K. Amyotrophic lateral sclerosis after accidental injection of mercury. J Neurol Neurosurg Psychiatry 1996; 60(6): 698.
[http://dx.doi.org/10.1136/jnnp.60.6.698] [PMID: 8648348]

[55] Adams CR, Ziegler DK, Lin JT. Mercury intoxication simulating amyotrophic lateral sclerosis. JAMA 1983; 250(5): 642-3.
[http://dx.doi.org/10.1001/jama.1983.03340050054029] [PMID: 6864963]

[56] Ngim C, Ngim AD. Health and safety in the dental clinic - Hygiene regulations for use of elemental mercury in the protection of rights, safety and well-being of the patients, workers and the environment. Singapore Dent J 2013; 34(1): 19-24.
[http://dx.doi.org/10.1016/j.sdj.2013.11.004] [PMID: 24360262]

[57]   Berlin M, Zalups RK, Fowler BA, Mercury GNordberg, Fowler B, Nordberg M. Handbook on the Toxicology of Metals 4ᵗʰ ed, volume II. United States: Academic Press. 2015; pp. 1013-75.

[58]   Woods JS, Heyer NJ, Russo JE, Martin MD, Farin FM. Genetic polymorphisms affecting susceptibility to mercury neurotoxicity in children: summary findings from the Casa Pia Children's Amalgam clinical trial. Neurotoxicology 2014; 44: 288-302.
[http://dx.doi.org/10.1016/j.neuro.2014.07.010] [PMID: 25109824]

[59]   Mutter J. Is dental amalgam safe for humans? The opinion of the scientific committee of the European Commission. J Occup Med Toxicol 2011; 6(1): 2.
[http://dx.doi.org/10.1186/1745-6673-6-2] [PMID: 21232090]

[60]   Attar AM, Kharkhaneh A, Etemadifar M, Keyhanian K, Davoudi V, Saadatnia M. Serum mercury level and multiple sclerosis. Biol Trace Elem Res 2012; 146(2): 150-3.
[http://dx.doi.org/10.1007/s12011-011-9239-y] [PMID: 22068727]

[61]   Kern JK, Geier DA, Deth RC, *et al.* Systematic assessment of research on autism spectrum disorder and mercury reveals conflicts of interest and the need for transparency in autism research. Sci Eng Ethics 2017; 23(6): 1691-718.
[http://dx.doi.org/10.1007/s11948-017-9983-2] [PMID: 29119411]

[62]   Skalny AV, Skalnaya MG, Bjørklund G, Nikonorov AA, Tinkov AA. Mercury as a possible link between maternal obesity and autism spectrum disorder. Med Hypotheses 2016; 91: 90-4.
[http://dx.doi.org/10.1016/j.mehy.2016.04.021] [PMID: 27142153]

[63]   Korbas M, O'Donoghue JL, Watson GE, *et al.* The chemical nature of mercury in human brain following poisoning or environmental exposure. ACS Chem Neurosci 2010; 1(12): 810-8.
[http://dx.doi.org/10.1021/cn1000765] [PMID: 22826746]

[64]   Mason LH, Harp JP, Han DY. Pb neurotoxicity: neuropsychological effects of lead toxicity. BioMed Res Int 2014; 2014: 840547.
[http://dx.doi.org/10.1155/2014/840547] [PMID: 24516855]

[65]   Struzyñska L, Bubko I, Walski M, Rafałowska U. Astroglial reaction during the early phase of acute lead toxicity in the adult rat brain. Toxicology 2001; 165(2-3): 121-31.
[http://dx.doi.org/10.1016/S0300-483X(01)00415-2] [PMID: 11522370]

[66]   Audesirk G. Electrophysiology of lead intoxication: effects on voltage-sensitive ion channels. Neurotoxicology 1993; 14(2-3): 137-47.
[PMID: 8247389]

[67]   Marchetti C. Molecular targets of lead in brain neurotoxicity. Neurotox Res 2003; 5(3): 221-36.
[http://dx.doi.org/10.1007/BF03033142] [PMID: 12835126]

[68]   Neal AP, Guilarte TR. Mechanisms of lead and manganese neurotoxicity. Toxicol Res (Camb) 2013; 2(2): 99-114.
[http://dx.doi.org/10.1039/c2tx20064c] [PMID: 25722848]

[69]   Olympio KP, Gonçalves C, Günther WM, Bechara EJ. Neurotoxicity and aggressiveness triggered by low-level lead in children: a review. Rev Panam Salud Publica 2009; 26(3): 266-75.
[http://dx.doi.org/10.1590/S1020-49892009000900011] [PMID: 20058837]

[70]   Nava-Ruiz C, Méndez-Armenta M, Ríos C. Lead neurotoxicity: effects on brain nitric oxide synthase. J Mol Histol 2012; 43(5): 553-63.
[http://dx.doi.org/10.1007/s10735-012-9414-2] [PMID: 22526509]

[71]   Ronchetti R, van den Hazel P, Schoeters G, *et al.* Lead neurotoxicity in children: is prenatal exposure more important than postnatal exposure? Acta Paediatr Suppl 2006; 95(453): 45-9.
[http://dx.doi.org/10.1111/j.1651-2227.2007.00144.x] [PMID: 17000569]

[72]   Bellinger DC. Lead neurotoxicity in children: decomposing the variability in dose-effect relationships. Am J Ind Med 2007; 50(10): 720-8.

[http://dx.doi.org/10.1002/ajim.20438] [PMID: 17290364]

[73]   Walkowiak J, Altmann L, Krämer U, *et al.* Cognitive and sensorimotor functions in 6-year-old children in relation to lead and mercury levels: adjustment for intelligence and contrast sensitivity in computerized testing. Neurotoxicol Teratol 1998; 20(5): 511-21.
[http://dx.doi.org/10.1016/S0892-0362(98)00010-5] [PMID: 9761589]

[74]   Bellinger DC. The protean toxicities of lead: new chapters in a familiar story. Int J Environ Res Public Health 2011; 8(7): 2593-628.
[http://dx.doi.org/10.3390/ijerph8072593] [PMID: 21845148]

[75]   Roy A, Bellinger D, Hu H, *et al.* Lead exposure and behavior among young children in Chennai, India. Environ Health Perspect 2009; 117(10): 1607-11.
[http://dx.doi.org/10.1289/ehp.0900625] [PMID: 20019913]

[76]   Dietrich KN, Berger OG, Succop PA. Lead exposure and the motor developmental status of urban six-year-old children in the Cincinnati Prospective Study. Pediatrics 1993; 91(2): 301-7.
[PMID: 7678702]

[77]   Sanders T, Liu Y, Buchner V, Tchounwou PB. Neurotoxic effects and biomarkers of lead exposure: a review. Rev Environ Health 2009; 24(1): 15-45.
[http://dx.doi.org/10.1515/REVEH.2009.24.1.15] [PMID: 19476290]

[78]   Toxicological profile for lead. Washington, DC, USA: US Department of Health and Human Services 2007.

[79]   Stewart WF, Schwartz BS, Davatzikos C, *et al.* Past adult lead exposure is linked to neurodegeneration measured by brain MRI. Neurology 2006; 66(10): 1476-84.
[http://dx.doi.org/10.1212/01.wnl.0000216138.69777.15] [PMID: 16717205]

[80]   Zhang H, Reynolds M. Cadmium exposure in living organisms: A short review. Sci Total Environ 2019; 678: 761-7.
[http://dx.doi.org/10.1016/j.scitotenv.2019.04.395] [PMID: 31085492]

[81]   Robards K, Worsfold P. Cadmium: toxicology and analysis. A review. Analyst (Lond) 1991; 116(6): 549-68.
[http://dx.doi.org/10.1039/an9911600549] [PMID: 1928728]

[82]   Järup L, Berglund M, Elinder CG, Nordberg G, Vahter M. Health effects of cadmium exposure--a review of the literature and a risk estimate. Scand J Work Environ Health 1998; 24 (Suppl. 1): 1-51.
[PMID: 9569444]

[83]   World Health Organization. Preventing disease through healthy environments: exposure to cadmium: a major public health concern (No WHO/CED/PHE/EPE/194 3) 2019.

[84]   Ganguly K, Levänen B, Palmberg L, Åkesson A, Lindén A. Cadmium in tobacco smokers: a neglected link to lung disease? Eur Respir Rev 2018; 27(147): 170122.
[http://dx.doi.org/10.1183/16000617.0122-2017] [PMID: 29592863]

[85]   Branca JJV, Morucci G, Pacini A. Cadmium-induced neurotoxicity: still much ado. Neural Regen Res 2018; 13(11): 1879-82.
[http://dx.doi.org/10.4103/1673-5374.239434] [PMID: 30233056]

[86]   Cannino G, Ferruggia E, Luparello C, Rinaldi AM. Cadmium and mitochondria. Mitochondrion 2009; 9(6): 377-84.
[http://dx.doi.org/10.1016/j.mito.2009.08.009] [PMID: 19706341]

[87]   Méndez-Armenta M, Ríos C. Cadmium neurotoxicity. Environ Toxicol Pharmacol 2007; 23(3): 350-8.
[http://dx.doi.org/10.1016/j.etap.2006.11.009] [PMID: 21783780]

[88]   Batool Z, Agha F, Tabassum S, Batool TS, Siddiqui RA, Haider S. Prevention of cadmium-induced neurotoxicity in rats by essential nutrients present in nuts. Acta Neurobiol Exp (Warsz) 2019; 79(2): 169-83.

[http://dx.doi.org/10.21307/ane-2019-015] [PMID: 31342953]

[89] López E, Arce C, Oset-Gasque MJ, Cañadas S, González MP. Cadmium induces reactive oxygen species generation and lipid peroxidation in cortical neurons in culture. Free Radic Biol Med 2006; 40(6): 940-51.
[http://dx.doi.org/10.1016/j.freeradbiomed.2005.10.062] [PMID: 16540389]

[90] Viaene MK, Masschelein R, Leenders J, De Groof M, Swerts LJ, Roels HA. Neurobehavioural effects of occupational exposure to cadmium: a cross sectional epidemiological study. Occup Environ Med 2000; 57(1): 19-27.
[http://dx.doi.org/10.1136/oem.57.1.19] [PMID: 10711265]

[91] Hart RP, Rose CS, Hamer RM. Neuropsychological effects of occupational exposure to cadmium. J Clin Exp Neuropsychol 1989; 11(6): 933-43.
[http://dx.doi.org/10.1080/01688638908400946] [PMID: 2592532]

[92] Tian LL, Zhao YC, Wang XC, *et al.* Effects of gestational cadmium exposure on pregnancy outcome and development in the offspring at age 4.5 years. Biol Trace Elem Res 2009; 132(1-3): 51-9.
[http://dx.doi.org/10.1007/s12011-009-8391-0] [PMID: 19404590]

[93] Kippler M, Tofail F, Hamadani JD, *et al.* Early-life cadmium exposure and child development in 5-year-old girls and boys: a cohort study in rural Bangladesh. Environ Health Perspect 2012; 120(10): 1462-8.
[http://dx.doi.org/10.1289/ehp.1104431] [PMID: 22759600]

[94] Jeong KS, Park H, Ha E, *et al.* Performance IQ in children is associated with blood cadmium concentration in early pregnancy. J Trace Elem Med Biol 2015; 30: 107-11.
[http://dx.doi.org/10.1016/j.jtemb.2014.11.007] [PMID: 25511909]

[95] Thatcher RW, Lester ML, McAlaster R, Horst R. Effects of low levels of cadmium and lead on cognitive functioning in children. Arch Environ Health 1982; 37(3): 159-66.
[http://dx.doi.org/10.1080/00039896.1982.10667557] [PMID: 7092333]

[96] Kilburn KH, McKinley KL. Persistent neurotoxicity from a battery fire: is cadmium the culprit? South Med J 1996; 89(7): 693-8.
[http://dx.doi.org/10.1097/00007611-199607000-00009] [PMID: 8685756]

[97] Min JY, Min KB. Blood cadmium levels and Alzheimer's disease mortality risk in older US adults. Environ Health 2016; 15(1): 69.
[http://dx.doi.org/10.1186/s12940-016-0155-7] [PMID: 27301955]

[98] Okuda B, Iwamoto Y, Tachibana H, Sugita M. Parkinsonism after acute cadmium poisoning. Clin Neurol Neurosurg 1997; 99(4): 263-5.
[http://dx.doi.org/10.1016/S0303-8467(97)00090-5] [PMID: 9491302]

[99] Bar-Sela S, Reingold S, Richter ED. Amyotrophic lateral sclerosis in a battery-factory worker exposed to cadmium. Int J Occup Environ Health 2001; 7(2): 109-12.
[http://dx.doi.org/10.1179/oeh.2001.7.2.109] [PMID: 11373040]

[100] Crichton R, Crichton RR, Boelaert JR. Inorganic biochemistry of iron metabolism: from molecular mechanisms to clinical consequences. United States: John Wiley & Sons 2001; p. 461.
[http://dx.doi.org/10.1002/0470845791]

[101] Ward RJ, Zucca FA, Duyn JH, Crichton RR, Zecca L. The role of iron in brain ageing and neurodegenerative disorders. Lancet Neurol 2014; 13(10): 1045-60.
[http://dx.doi.org/10.1016/S1474-4422(14)70117-6] [PMID: 25231526]

[102] Schipper HM. Brain iron deposition and the free radical-mitochondrial theory of ageing. Ageing Res Rev 2004; 3(3): 265-301.
[http://dx.doi.org/10.1016/j.arr.2004.02.001] [PMID: 15231237]

[103] Gutteridge JM. Iron and oxygen radicals in brain. Ann Neurol 1992; 32 (Suppl.): S16-21.
[http://dx.doi.org/10.1002/ana.410320705] [PMID: 1510375]

[104] Hallgren B, Sourander P. The effect of age on the non-haemin iron in the human brain. J Neurochem 1958; 3(1): 41-51.
[http://dx.doi.org/10.1111/j.1471-4159.1958.tb12607.x] [PMID: 13611557]

[105] Thomas LO, Boyko OB, Anthony DC, Burger PC. MR detection of brain iron. AJNR Am J Neuroradiol 1993; 14(5): 1043-8.
[PMID: 8237678]

[106] Drayer B, Burger P, Darwin R, Riederer S, Herfkens R, Johnson GA. MRI of brain iron. AJR Am J Roentgenol 1986; 147(1): 103-10.
[http://dx.doi.org/10.2214/ajr.147.1.103] [PMID: 3487201]

[107] Bartzokis G, Tishler TA, Lu PH, *et al.* Brain ferritin iron may influence age- and gender-related risks of neurodegeneration. Neurobiol Aging 2007; 28(3): 414-23.
[http://dx.doi.org/10.1016/j.neurobiolaging.2006.02.005] [PMID: 16563566]

[108] Bartzokis G, Beckson M, Hance DB, Marx P, Foster JA, Marder SR. MR evaluation of age-related increase of brain iron in young adult and older normal males. Magn Reson Imaging 1997; 15(1): 29-35.
[http://dx.doi.org/10.1016/S0730-725X(96)00234-2] [PMID: 9084022]

[109] Stankiewicz J, Panter SS, Neema M, Arora A, Batt CE, Bakshi R. Iron in chronic brain disorders: imaging and neurotherapeutic implications. Neurotherapeutics 2007; 4(3): 371-86.
[http://dx.doi.org/10.1016/j.nurt.2007.05.006] [PMID: 17599703]

[110] Pujol J, Junqué C, Vendrell P, *et al.* Biological significance of iron-related magnetic resonance imaging changes in the brain. Arch Neurol 1992; 49(7): 711-7.
[http://dx.doi.org/10.1001/archneur.1992.00530310053012] [PMID: 1497497]

[111] Ritchie CW, Bush AI, Mackinnon A, *et al.* Metal-protein attenuation with iodochlorhydroxyquin (clioquinol) targeting Abeta amyloid deposition and toxicity in Alzheimer disease: a pilot phase 2 clinical trial. Arch Neurol 2003; 60(12): 1685-91.
[http://dx.doi.org/10.1001/archneur.60.12.1685] [PMID: 14676042]

[112] Berg D, Hochstrasser H. Iron metabolism in Parkinsonian syndromes. Mov Disord 2006; 21(9): 1299-310.
[http://dx.doi.org/10.1002/mds.21020] [PMID: 16817199]

[113] Youdim MB, Ben-Shachar D, Riederer P. Is Parkinson's disease a progressive siderosis of substantia nigra resulting in iron and melanin induced neurodegeneration? Acta Neurol Scand Suppl 1989; 126 (Suppl.): 47-54.
[http://dx.doi.org/10.1111/j.1600-0404.1989.tb01782.x] [PMID: 2618593]

[114] Zecca L, Zucca FA, Albertini A, Rizzio E, Fariello RG. A proposed dual role of neuromelanin in the pathogenesis of Parkinson's disease. Neurology 2006; 67(7) (Suppl. 2): S8-S11.
[http://dx.doi.org/10.1212/WNL.67.7_suppl_2.S8] [PMID: 17030740]

[115] Bartzokis G, Cummings JL, Markham CH, *et al.* MRI evaluation of brain iron in earlier- and later-onset Parkinson's disease and normal subjects. Magn Reson Imaging 1999; 17(2): 213-22.
[http://dx.doi.org/10.1016/S0730-725X(98)00155-6] [PMID: 10215476]

[116] Levenson CW. Iron and Parkinson's disease: chelators to the rescue? Nutr Rev 2003; 61(9): 311-3.
[http://dx.doi.org/10.1301/nr.2003.sept.311-313] [PMID: 14552066]

[117] Dexter DT, Carayon A, Vidailhet M, *et al.* Decreased ferritin levels in brain in Parkinson's disease. J Neurochem 1990; 55(1): 16-20.
[http://dx.doi.org/10.1111/j.1471-4159.1990.tb08814.x] [PMID: 2355217]

[118] Zecca L, Berg D, Arzberger T, *et al.* In vivo detection of iron and neuromelanin by transcranial sonography: a new approach for early detection of substantia nigra damage. Mov Disord 2005; 20(10): 1278-85.
[http://dx.doi.org/10.1002/mds.20550] [PMID: 15986424]

[119] Devos D, Moreau C, Devedjian JC, *et al.* Targeting chelatable iron as a therapeutic modality in Parkinson's disease. Antioxid Redox Signal 2014; 21(2): 195-210.
[http://dx.doi.org/10.1089/ars.2013.5593] [PMID: 24251381]

[120] Grolez G, Moreau C, Sablonnière B, *et al.* Ceruloplasmin activity and iron chelation treatment of patients with Parkinson's disease. BMC Neurol 2015; 15: 74.
[http://dx.doi.org/10.1186/s12883-015-0331-3] [PMID: 25943368]

[121] Bakshi R, Shaikh ZA, Janardhan V. MRI T2 shortening ('black T2') in multiple sclerosis: frequency, location, and clinical correlation. Neuroreport 2000; 11(1): 15-21.
[http://dx.doi.org/10.1097/00001756-200001170-00004] [PMID: 10683822]

[122] Bakshi R, Benedict RH, Bermel RA, *et al.* T2 hypointensity in the deep gray matter of patients with multiple sclerosis: a quantitative magnetic resonance imaging study. Arch Neurol 2002; 59(1): 62-8.
[http://dx.doi.org/10.1001/archneur.59.1.62] [PMID: 11790232]

[123] Ge Y, Jensen JH, Lu H, *et al.* Quantitative assessment of iron accumulation in the deep gray matter of multiple sclerosis by magnetic field correlation imaging. AJNR Am J Neuroradiol 2007; 28(9): 1639-44.
[http://dx.doi.org/10.3174/ajnr.A0646] [PMID: 17893225]

[124] Tjoa CW, Benedict RH, Weinstock-Guttman B, Fabiano AJ, Bakshi R. MRI T2 hypointensity of the dentate nucleus is related to ambulatory impairment in multiple sclerosis. J Neurol Sci 2005; 234(1-2): 17-24.
[http://dx.doi.org/10.1016/j.jns.2005.02.009] [PMID: 15993137]

[125] Neema M, Stankiewicz J, Arora A, *et al.* T1- and T2-based MRI measures of diffuse gray matter and white matter damage in patients with multiple sclerosis. J Neuroimaging 2007; 17 (Suppl. 1): 16S-21S.
[http://dx.doi.org/10.1111/j.1552-6569.2007.00131.x] [PMID: 17425729]

[126] Drayer B, Burger P, Hurwitz B, Dawson D, Cain J. Reduced signal intensity on MR images of thalamus and putamen in multiple sclerosis: increased iron content? AJR Am J Roentgenol 1987; 149(2): 357-63.
[http://dx.doi.org/10.2214/ajr.149.2.357] [PMID: 3496764]

[127] Craelius W, Migdal MW, Luessenhop CP, Sugar A, Mihalakis I. Iron deposits surrounding multiple sclerosis plaques. Arch Pathol Lab Med 1982; 106(8): 397-9.
[PMID: 6896630]

[128] LeVine SM. Iron deposits in multiple sclerosis and Alzheimer's disease brains. Brain Res 1997; 760(1-2): 298-303.
[http://dx.doi.org/10.1016/S0006-8993(97)00470-8] [PMID: 9237552]

[129] Bowern N, Ramshaw IA, Clark IA, Doherty PC. Inhibition of autoimmune neuropathological process by treatment with an iron-chelating agent. J Exp Med 1984; 160(5): 1532-43.
[http://dx.doi.org/10.1084/jem.160.5.1532] [PMID: 6333485]

[130] Willenborg DO, Bowern NA, Danta G, Doherty PC. Inhibition of allergic encephalomyelitis by the iron chelating agent desferrioxamine: differential effect depending on type of sensitizing encephalitogen. J Neuroimmunol 1988; 17(2): 127-35.
[http://dx.doi.org/10.1016/0165-5728(88)90020-3] [PMID: 2447124]

[131] Anderson RA. Nutritional role of chromium. Sci Total Environ 1981; 17(1): 13-29.
[http://dx.doi.org/10.1016/0048-9697(81)90104-2] [PMID: 7010598]

[132] Pechova A, Pavlata L. Chromium as an essential nutrient: a review. Vet Med (Prague, Czech Repub) 2007; 52: 1-18.

[133] Jain SK, Rains JL, Croad JL. Effect of chromium niacinate and chromium picolinate supplementation on lipid peroxidation, TNF-α, IL-6, CRP, glycated hemoglobin, triglycerides, and cholesterol levels in blood of streptozotocin-treated diabetic rats. Free Radic Biol Med 2007; 43(8): 1124-31.
[http://dx.doi.org/10.1016/j.freeradbiomed.2007.05.019] [PMID: 17854708]

[134] Vincent JB. Recent developments in the biochemistry of chromium(III). Biol Trace Elem Res 2004; 99(1-3): 1-16.
[http://dx.doi.org/10.1385/BTER:99:1-3:001] [PMID: 15235137]

[135] Cefalu WT, Wang ZQ, Zhang XH, Baldor LC, Russell JC. Oral chromium picolinate improves carbohydrate and lipid metabolism and enhances skeletal muscle Glut-4 translocation in obese, hyperinsulinemic (JCR-LA corpulent) rats. J Nutr 2002; 132(6): 1107-14.
[http://dx.doi.org/10.1093/jn/132.6.1107] [PMID: 12042418]

[136] Guertin J. Toxicity and health effects of chromium (all oxidation states). Chromium (VI) handbook 2004; 215-34.

[137] Miesel R, Kröger H, Kurpisz M, Weser U. Induction of arthritis in mice and rats by potassium peroxochromate and assessment of disease activity by whole blood chemiluminescence and 99mpertechnetate-imaging. Free Radic Res 1995; 23(3): 213-27.
[http://dx.doi.org/10.3109/10715769509064035] [PMID: 7581817]

[138] Travacio M, Polo JM, Llesuy S. Chromium (VI) induces oxidative stress in the mouse brain. Toxicology 2001; 162(2): 139-48.
[http://dx.doi.org/10.1016/S0300-483X(00)00423-6] [PMID: 11337112]

[139] Fu SC, Liu JM, Lee KI, *et al.* Cr(VI) induces ROS-mediated mitochondrial-dependent apoptosis in neuronal cells *via* the activation of Akt/ERK/AMPK signaling pathway. Toxicol In Vitro 2020; 65: 104795.
[http://dx.doi.org/10.1016/j.tiv.2020.104795] [PMID: 32061800]

[140] Vingtdeux V, Davies P, Dickson DW, Marambaud P. AMPK is abnormally activated in tangle- and pre-tangle-bearing neurons in Alzheimer's disease and other tauopathies. Acta Neuropathol 2011; 121(3): 337-49.
[http://dx.doi.org/10.1007/s00401-010-0759-x] [PMID: 20957377]

[141] Chou SY, Lee YC, Chen HM, *et al.* CGS21680 attenuates symptoms of Huntington's disease in a transgenic mouse model. J Neurochem 2005; 93(2): 310-20.
[http://dx.doi.org/10.1111/j.1471-4159.2005.03029.x] [PMID: 15816854]

[142] Xu Y, Liu C, Chen S, *et al.* Activation of AMPK and inactivation of Akt result in suppression of mTOR-mediated S6K1 and 4E-BP1 pathways leading to neuronal cell death in *in vitro* models of Parkinson's disease. Cell Signal 2014; 26(8): 1680-9.
[http://dx.doi.org/10.1016/j.cellsig.2014.04.009] [PMID: 24726895]

[143] Lieberman H. Chrome Ulcerations of the Nose and Throat. N Engl J Med 1941; 225(4): 132-3.
[http://dx.doi.org/10.1056/NEJM194107242250402]

[144] Wecker L, Miller SB, Cochran SR, Dugger DL, Johnson WD. Trace element concentrations in hair from autistic children. J Ment Defic Res 1985; 29(Pt 1): 15-22.
[http://dx.doi.org/10.1111/j.1365-2788.1985.tb00303.x] [PMID: 4009700]

[145] Adams JB, Holloway CE, George F, Quig D. Analyses of toxic metals and essential minerals in the hair of Arizona children with autism and associated conditions, and their mothers. Biol Trace Elem Res 2006; 110(3): 193-209.
[http://dx.doi.org/10.1385/BTER:110:3:193] [PMID: 16845157]

[146] Yorbik O, Kurt I, Haşimi A, Oztürk O. Chromium, cadmium, and lead levels in urine of children with autism and typically developing controls. Biol Trace Elem Res 2010; 135(1-3): 10-5.
[http://dx.doi.org/10.1007/s12011-009-8494-7] [PMID: 19688188]

[147] Vojdani A, Mumper E, Granpeesheh D, *et al.* Low natural killer cell cytotoxic activity in autism: the role of glutathione, IL-2 and IL-15. J Neuroimmunol 2008; 205(1-2): 148-54.
[http://dx.doi.org/10.1016/j.jneuroim.2008.09.005] [PMID: 18929414]

[148] Geier DA, Geier MR. A clinical and laboratory evaluation of methionine cycle-transsulfuration and androgen pathway markers in children with autistic disorders. Horm Res 2006; 66(4): 182-8.

[PMID: 16825783]

[149] Geier DA, Kern JK, Garver CR, Adams JB, Audhya T, Geier MR. A prospective study of transsulfuration biomarkers in autistic disorders. Neurochem Res 2009; 34(2): 386-93.
[http://dx.doi.org/10.1007/s11064-008-9782-x] [PMID: 18612812]

[150] James SJ, Melnyk S, Fuchs G, *et al.* Efficacy of methylcobalamin and folinic acid treatment on glutathione redox status in children with autism. Am J Clin Nutr 2009; 89(1): 425-30.
[http://dx.doi.org/10.3945/ajcn.2008.26615] [PMID: 19056591]

[151] Duckett S. Abnormal deposits of chromium in the pathological human brain. J Neurol Neurosurg Psychiatry 1986; 49(3): 296-301.
[http://dx.doi.org/10.1136/jnnp.49.3.296] [PMID: 3958742]

[152] Schrauzer GN, Shrestha KP, Flores-Arce M. Somatopsychological effects of chromium supplementation. J Nutr Med 1992; 3: 43-8.
[http://dx.doi.org/10.3109/13590849208997960]

[153] Lovrincevic I, Leung FY, Alfieri MA, Grace DM. Can elevated chromium induce somatopsychic responses? Biol Trace Elem Res 1996; 55(1-2): 163-71.
[http://dx.doi.org/10.1007/BF02784177] [PMID: 8971363]

[154] Clark MJ, Prentice JR, Hoggard N, Paley MN, Hadjivassiliou M, Wilkinson JM. Brain structure and function in patients after metal-on-metal hip resurfacing. AJNR Am J Neuroradiol 2014; 35(9): 1753-8.
[http://dx.doi.org/10.3174/ajnr.A3922] [PMID: 24722312]

[155] Huk OL, Catelas I, Mwale F, Antoniou J, Zukor DJ, Petit A. Induction of apoptosis and necrosis by metal ions *in vitro.* J Arthroplasty 2004; 19(8) (Suppl. 3): 84-7.
[http://dx.doi.org/10.1016/j.arth.2004.09.011] [PMID: 15578559]

[156] Quinteros FA, Poliandri AH, Machiavelli LI, Cabilla JP, Duvilanski BH. In vivo and *in vitro* effects of chromium VI on anterior pituitary hormone release and cell viability. Toxicol Appl Pharmacol 2007; 218(1): 79-87.
[http://dx.doi.org/10.1016/j.taap.2006.10.017] [PMID: 17141818]

[157] Karovic O, Tonazzini I, Rebola N, *et al.* Toxic effects of cobalt in primary cultures of mouse astrocytes. Similarities with hypoxia and role of HIF-1alpha. Biochem Pharmacol 2007; 73(5): 694-708.
[http://dx.doi.org/10.1016/j.bcp.2006.11.008] [PMID: 17169330]

[158] Prentice JR, Blackwell CS, Raoof N, *et al.* Auditory and visual health after ten years of exposure to metal-on-metal hip prostheses: a cross-sectional study follow up. PLoS One 2014; 9(3): e90838.
[http://dx.doi.org/10.1371/journal.pone.0090838] [PMID: 24621561]

[159] Rizzetti MC, Liberini P, Zarattini G, *et al.* Loss of sight and sound. Could it be the hip? Lancet 2009; 373(9668): 1052.
[http://dx.doi.org/10.1016/S0140-6736(09)60490-6] [PMID: 19304018]

[160] Ikeda T, Takahashi K, Kabata T, Sakagoshi D, Tomita K, Yamada M. Polyneuropathy caused by cobalt-chromium metallosis after total hip replacement. Muscle Nerve 2010; 42(1): 140-3.
[http://dx.doi.org/10.1002/mus.21638] [PMID: 20544916]

[161] Green B, Griffiths E, Almond S. Neuropsychiatric symptoms following metal-on-metal implant failure with cobalt and chromium toxicity. BMC Psychiatry 2017; 17(1): 33.
[http://dx.doi.org/10.1186/s12888-016-1174-1] [PMID: 28114963]

[162] Caparros-Gonzalez RA, Giménez-Asensio MJ, González-Alzaga B, *et al.* Childhood chromium exposure and neuropsychological development in children living in two polluted areas in southern Spain. Environ Pollut 2019; 252((Pt B)): 1550-60.
[http://dx.doi.org/10.1016/j.envpol.2019.06.084]

[163] Varela-Moreiras G, Murphy MM, Scott JM. Cobalamin, folic acid, and homocysteine. Nutr Rev 2009;

67 (Suppl. 1): S69-72.
[http://dx.doi.org/10.1111/j.1753-4887.2009.00163.x] [PMID: 19453682]

[164] Lison D. Cobalt.Handbook on the toxicology of metals. 3rd ed. Burlington, USA: Elsevier 2007; pp. 511-28.
[http://dx.doi.org/10.1016/B978-012369413-3/50080-X]

[165] Barceloux DG. Cobalt. J Toxicol Clin Toxicol 1999; 37(2): 201-6.
[http://dx.doi.org/10.1081/CLT-100102420] [PMID: 10382556]

[166] Scansetti G, Botta GC, Spinelli P, Reviglione L, Ponzetti C. Absorption and excretion of cobalt in the hard metal industry. Sci Total Environ 1994; 150(1-3): 141-4.
[http://dx.doi.org/10.1016/0048-9697(94)90141-4] [PMID: 7939587]

[167] Lan A, Liao X, Mo L, *et al.* Hydrogen sulfide protects against chemical hypoxia-induced injury by inhibiting ROS-activated ERK1/2 and p38MAPK signaling pathways in PC12 cells. PLoS One 2011; 6(10): e25921.
[http://dx.doi.org/10.1371/journal.pone.0025921] [PMID: 21998720]

[168] Zou W, Zeng J, Zhuo M, *et al.* Involvement of caspase-3 and p38 mitogen-activated protein kinase in cobalt chloride-induced apoptosis in PC12 cells. J Neurosci Res 2002; 67(6): 837-43.
[http://dx.doi.org/10.1002/jnr.10168] [PMID: 11891799]

[169] Zhong X, Lin R, Li Z, Mao J, Chen L. Effects of Salidroside on cobalt chloride-induced hypoxia damage and mTOR signaling repression in PC12 cells. Biol Pharm Bull 2014; 37(7): 1199-206.
[http://dx.doi.org/10.1248/bpb.b14-00100] [PMID: 24989011]

[170] Lan AP, Chen J, Chai ZF, Hu Y. The neurotoxicity of iron, copper and cobalt in Parkinson's disease through ROS-mediated mechanisms. Biometals 2016; 29(4): 665-78.
[http://dx.doi.org/10.1007/s10534-016-9942-4] [PMID: 27349232]

[171] Jordan C, Whitman RD, Harbut M, Tanner B. Memory deficits in workers suffering from hard metal disease. Toxicol Lett 1990; 54(2-3): 241-3.
[http://dx.doi.org/10.1016/0378-4274(90)90190-W] [PMID: 2260122]

[172] Meecham HM, Humphrey P. Industrial exposure to cobalt causing optic atrophy and nerve deafness: a case report. J Neurol Neurosurg Psychiatry 1991; 54(4): 374-5.
[http://dx.doi.org/10.1136/jnnp.54.4.374] [PMID: 2056332]

[173] Domingo JL. Cobalt in the environment and its toxicological implications. Rev Environ Contam Toxicol 1989; 108: 105-32.
[http://dx.doi.org/10.1007/978-1-4613-8850-0_3] [PMID: 2646660]

[174] Gardner FH. The use of cobaltous chloride in the anemia associated with chronic renal disease. J Lab Clin Med 1953; 41(1): 56-64.
[PMID: 13023095]

[175] Schirrmacher UOE. Case of cobalt poisoning. BMJ 1967; 1(5539): 544-5.
[http://dx.doi.org/10.1136/bmj.1.5539.544] [PMID: 6017158]

[176] Licht A, Oliver M, Rachmilewitz EA. Optic atrophy following treatment with cobalt chloride in a patient with pancytopenia and hypercellular marrow. Isr J Med Sci 1972; 8(1): 61-6.
[PMID: 5026863]

[177] Megaterio S, Galetto F, Alossa E, Capretto S. Systemic effects of ionic release in wear of prosthetic head: a case report. Giorn It Ort Traum 2001; 27: 173-5.

[178] Steens W, von Foerster G, Katzer A. Severe cobalt poisoning with loss of sight after ceramic-metal pairing in a hip--a case report. Acta Orthop 2006; 77(5): 830-2.
[http://dx.doi.org/10.1080/17453670610013079] [PMID: 17068719]

[179] Oldenburg M, Wegner R, Baur X. Severe cobalt intoxication due to prosthesis wear in repeated total hip arthroplasty. J Arthroplasty 2009; 24(5): 825.e15-20.

[http://dx.doi.org/10.1016/j.arth.2008.07.017] [PMID: 18835128]

[180] Tower SS. Arthroprosthetic cobaltism: neurological and cardiac manifestations in two patients with metal-on-metal arthroplasty: a case report. J Bone Joint Surg Am 2010; 92(17): 2847-51.
[http://dx.doi.org/10.2106/JBJS.J.00125] [PMID: 21037026]

[181] Catalani S, Rizzetti MC, Padovani A, Apostoli P. Neurotoxicity of cobalt. Hum Exp Toxicol 2012; 31(5): 421-37.
[http://dx.doi.org/10.1177/0960327111414280] [PMID: 21729976]

[182] Solomons NW. Biochemical, metabolic, and clinical role of copper in human nutrition. J Am Coll Nutr 1985; 4(1): 83-105.
[http://dx.doi.org/10.1080/07315724.1985.10720069] [PMID: 3921587]

[183] Trace Elements in Human Nutrition and Health. Geneva: World Health Organization 1996.

[184] Mills CF. The physiological roles of copper. Food Chem 1992; 3(3): 239-40.
[http://dx.doi.org/10.1016/0308-8146(92)90180-A]

[185] Percival SS, Harris ED. Regulation of Cu,Zn superoxide dismutase with copper. Caeruloplasmin maintains levels of functional enzyme activity during differentiation of K562 cells. Biochem J 1991; 274(Pt 1): 153-8.
[http://dx.doi.org/10.1042/bj2740153] [PMID: 1900417]

[186] Abou Zeid C, Kaler SG. Normal Human Copper Metabolism.Wilson Disease: Pathogenesis, molecular mechanisms, diagnosis, treatment and monitoring. 1st ed. United States: Academic Press 2019; pp. 17-22.
[http://dx.doi.org/10.1016/B978-0-12-811077-5.00002-5]

[187] Halliwell B, Gutteridge JM. Oxygen toxicity, oxygen radicals, transition metals and disease. Biochem J 1984; 219(1): 1-14.
[http://dx.doi.org/10.1042/bj2190001] [PMID: 6326753]

[188] Halliwell B, Gutteridge JM. Role of free radicals and catalytic metal ions in human disease: an overview. Methods Enzymol 1990; 186: 1-85.
[http://dx.doi.org/10.1016/0076-6879(90)86093-B] [PMID: 2172697]

[189] Madsen E, Gitlin JD. Copper and iron disorders of the brain. Annu Rev Neurosci 2007; 30: 317-37.
[http://dx.doi.org/10.1146/annurev.neuro.30.051606.094232] [PMID: 17367269]

[190] Gupta A, Lutsenko S. Human copper transporters: mechanism, role in human diseases and therapeutic potential. Future Med Chem 2009; 1(6): 1125-42.
[http://dx.doi.org/10.4155/fmc.09.84] [PMID: 20454597]

[191] Kaler SG, Holmes CS, Goldstein DS, et al. Neonatal diagnosis and treatment of Menkes disease. N Engl J Med 2008; 358(6): 605-14.
[http://dx.doi.org/10.1056/NEJMoa070613] [PMID: 18256395]

[192] Tümer Z, Møller LB. Menkes disease. Eur J Hum Genet 2010; 18(5): 511-8.
[http://dx.doi.org/10.1038/ejhg.2009.187] [PMID: 19888294]

[193] Kodama H, Fujisawa C, Bhadhprasit W. Pathology, clinical features and treatments of congenital copper metabolic disorders--focus on neurologic aspects. Brain Dev 2011; 33(3): 243-51.
[http://dx.doi.org/10.1016/j.braindev.2010.10.021] [PMID: 21112168]

[194] Taylor AA, Tsuji JS, Garry MR, et al. Critical Review of Exposure and Effects: Implications for Setting Regulatory Health Criteria for Ingested Copper. Environ Manage 2020; 65(1): 131-59.
[http://dx.doi.org/10.1007/s00267-019-01234-y] [PMID: 31832729]

[195] Desai V, Kaler SG. Role of copper in human neurological disorders. Am J Clin Nutr 2008; 88(3): 855S-8S.
[http://dx.doi.org/10.1093/ajcn/88.3.855S] [PMID: 18779308]

[196] Bagheri S, Squitti R, Haertlé T, Siotto M, Saboury AA. role of copper in the onset of Alzheimer's

disease compared to other metals. Front Aging Neurosci 2018; 9: 446.
[http://dx.doi.org/10.3389/fnagi.2017.00446] [PMID: 29472855]

[197] Dai XL, Sun YX, Jiang ZF. Cu(II) potentiation of Alzheimer Abeta1-40 cytotoxicity and transition on its secondary structure. Acta Biochim Biophys Sin (Shanghai) 2006; 38(11): 765-72.
[http://dx.doi.org/10.1111/j.1745-7270.2006.00228.x] [PMID: 17091193]

[198] Tõugu V, Karafin A, Zovo K, *et al.* Zn(II)- and Cu(II)-induced non-fibrillar aggregates of amyloid-beta (1-42) peptide are transformed to amyloid fibrils, both spontaneously and under the influence of metal chelators. J Neurochem 2009; 110(6): 1784-95.
[http://dx.doi.org/10.1111/j.1471-4159.2009.06269.x] [PMID: 19619132]

[199] Miller LM, Wang Q, Telivala TP, Smith RJ, Lanzirotti A, Miklossy J. Synchrotron-based infrared and X-ray imaging shows focalized accumulation of Cu and Zn co-localized with beta-amyloid deposits in Alzheimer's disease. J Struct Biol 2006; 155(1): 30-7.
[http://dx.doi.org/10.1016/j.jsb.2005.09.004] [PMID: 16325427]

[200] Lovell MA, Robertson JD, Teesdale WJ, Campbell JL, Markesbery WR. Copper, iron and zinc in Alzheimer's disease senile plaques. J Neurol Sci 1998; 158(1): 47-52.
[http://dx.doi.org/10.1016/S0022-510X(98)00092-6] [PMID: 9667777]

[201] James SA, Volitakis I, Adlard PA, *et al.* Elevated labile Cu is associated with oxidative pathology in Alzheimer disease. Free Radic Biol Med 2012; 52(2): 298-302.
[http://dx.doi.org/10.1016/j.freeradbiomed.2011.10.446] [PMID: 22080049]

[202] Szabo ST, Harry GJ, Hayden KM, Szabo DT, Birnbaum L. Comparison of metal levels between postmortem brain and ventricular fluid in Alzheimer's disease and non-demented elderly controls. Toxicol Sci 2016; 150(2): 292-300.
[http://dx.doi.org/10.1093/toxsci/kfv325] [PMID: 26721301]

[203] Schrag M, Mueller C, Oyoyo U, Smith MA, Kirsch WM. Iron, zinc and copper in the Alzheimer's disease brain: a quantitative meta-analysis. Some insight on the influence of citation bias on scientific opinion. Prog Neurobiol 2011; 94(3): 296-306.
[http://dx.doi.org/10.1016/j.pneurobio.2011.05.001] [PMID: 21600264]

[204] Bucossi S, Ventriglia M, Panetta V, *et al.* Copper in Alzheimer's disease: a meta-analysis of serum,plasma, and cerebrospinal fluid studies. J Alzheimers Dis 2011; 24(1): 175-85.
[http://dx.doi.org/10.3233/JAD-2010-101473] [PMID: 21187586]

[205] Ventriglia M, Bucossi S, Panetta V, Squitti R. Copper in Alzheimer's disease: a meta-analysis of serum, plasma, and cerebrospinal fluid studies. J Alzheimers Dis 2012; 30(4): 981-4.
[http://dx.doi.org/10.3233/JAD-2012-120244] [PMID: 22475798]

[206] Schrag M, Mueller C, Zabel M, *et al.* Oxidative stress in blood in Alzheimer's disease and mild cognitive impairment: a meta-analysis. Neurobiol Dis 2013; 59: 100-10.
[http://dx.doi.org/10.1016/j.nbd.2013.07.005] [PMID: 23867235]

[207] Squitti R, Simonelli I, Ventriglia M, *et al.* Meta-analysis of serum non-ceruloplasmin copper in Alzheimer's disease. J Alzheimers Dis 2014; 38(4): 809-22.
[http://dx.doi.org/10.3233/JAD-131247] [PMID: 24072069]

[208] Wang ZX, Tan L, Wang HF, *et al.* Serum Iron, Zinc, and Copper Levels in Patients with Alzheimer's Disease: A Replication Study and Meta-Analyses. J Alzheimers Dis 2015; 47(3): 565-81.
[http://dx.doi.org/10.3233/JAD-143108] [PMID: 26401693]

[209] Pu Z, Xu W, Lin Y, He J, Huang M. Oxidative stress markers and metal ions are correlated with cognitive function in Alzheimer's disease. Am J Alzheimers Dis Other Demen 2017; 32(6): 353-9.
[http://dx.doi.org/10.1177/1533317517709549] [PMID: 28554217]

[210] Talwar P, Grover S, Sinha J, *et al.* multifactorial analysis of a biomarker pool for Alzheimer disease risk in a North Indian population. Dement Geriatr Cogn Disord 2017; 44(1-2): 25-34.
[http://dx.doi.org/10.1159/000477206] [PMID: 28633142]

[211] Guan C, Dang R, Cui Y, *et al.* Characterization of plasma metal profiles in Alzheimer's disease using multivariate statistical analysis. PLoS One 2017; 12(7): e0178271.
[http://dx.doi.org/10.1371/journal.pone.0178271] [PMID: 28719622]

[212] Gorell JM, Johnson CC, Rybicki BA, *et al.* Occupational exposure to manganese, copper, lead, iron, mercury and zinc and the risk of Parkinson's disease. Neurotoxicology 1999; 20(2-3): 239-47.
[PMID: 10385887]

[213] Willis AW, Evanoff BA, Lian M, *et al.* Metal emissions and urban incident Parkinson disease: a community health study of Medicare beneficiaries by using geographic information systems. Am J Epidemiol 2010; 172(12): 1357-63.
[http://dx.doi.org/10.1093/aje/kwq303] [PMID: 20959505]

[214] Ahmed SS, Santosh W. Metallomic profiling and linkage map analysis of early Parkinson's disease: a new insight to Aluminium marker for the possible diagnosis. PLoS One 2010; 5(6): e11252.
[http://dx.doi.org/10.1371/journal.pone.0011252] [PMID: 20582167]

[215] Pall HS, Williams AC, Blake DR, *et al.* Raised cerebrospinal-fluid copper concentration in Parkinson's disease. Lancet 1987; 2(8553): 238-41.
[http://dx.doi.org/10.1016/S0140-6736(87)90827-0] [PMID: 2886715]

[216] Hozumi I, Hasegawa T, Honda A, *et al.* Patterns of levels of biological metals in CSF differ among neurodegenerative diseases. J Neurol Sci 2011; 303(1-2): 95-9.
[http://dx.doi.org/10.1016/j.jns.2011.01.003] [PMID: 21292280]

[217] Larner F, Sampson B, Rehkämper M, *et al.* High precision isotope measurements reveal poor control of copper metabolism in parkinsonism. Metallomics 2013; 5(2): 125-32.
[http://dx.doi.org/10.1039/c3mt20238k] [PMID: 23340956]

[218] Yeargin-Allsopp M, Rice C, Karapurkar T, Doernberg N, Boyle C, Murphy C. Prevalence of autism in a US metropolitan area. JAMA 2003; 289(1): 49-55.
[http://dx.doi.org/10.1001/jama.289.1.49] [PMID: 12503976]

[219] Muhle R, Trentacoste SV, Rapin I. The genetics of autism. Pediatrics 2004; 113(5): e472-86.
[http://dx.doi.org/10.1542/peds.113.5.e472] [PMID: 15121991]

[220] Russo AJ. Increased copper in individuals with autism normalizes post zinc therapy more efficiently in individuals with concurrent GI disease. Nutr Metab Insights 2011; 4: 49-54.
[http://dx.doi.org/10.4137/NMI.S6827] [PMID: 23946661]

[221] Vergani L, Lanza C, Rivaro P, *et al.* Metals, metallothioneins and oxidative stress in blood of autistic children. Res Autism Spectr Disord 2011; 5: 286-93.
[http://dx.doi.org/10.1016/j.rasd.2010.04.010]

[222] Dexter DT, Carayon A, Javoy-Agid F, *et al.* Alterations in the levels of iron, ferritin and other trace metals in Parkinson's disease and other neurodegenerative diseases affecting the basal ganglia. Brain 1991; 114(Pt 4): 1953-75.
[http://dx.doi.org/10.1093/brain/114.4.1953] [PMID: 1832073]

[223] Fox JH, Kama JA, Lieberman G, *et al.* Mechanisms of copper ion mediated Huntington's disease progression. PLoS One 2007; 2(3): e334.
[http://dx.doi.org/10.1371/journal.pone.0000334] [PMID: 17396163]

[224] Xiao G, Fan Q, Wang X, Zhou B. Huntington disease arises from a combinatory toxicity of polyglutamine and copper binding. Proc Natl Acad Sci USA 2013; 110(37): 14995-5000.
[http://dx.doi.org/10.1073/pnas.1308535110] [PMID: 23980182]

[225] Brown DR. Copper and prion diseases. Biochem Soc Trans 2002; 30(4): 742-5.
[http://dx.doi.org/10.1042/bst0300742] [PMID: 12196183]

[226] Sigurdsson EM, Brown DR, Alim MA, *et al.* Copper chelation delays the onset of prion disease. J Biol Chem 2003; 278(47): 46199-202.

[http://dx.doi.org/10.1074/jbc.C300303200] [PMID: 14519758]

[227] Aschner M, Guilarte TR, Schneider JS, Zheng W. Manganese: recent advances in understanding its transport and neurotoxicity. Toxicol Appl Pharmacol 2007; 221(2): 131-47.
[http://dx.doi.org/10.1016/j.taap.2007.03.001] [PMID: 17466353]

[228] Levy BS, Nassetta WJ. Neurologic effects of manganese in humans: a review. Int J Occup Environ Health 2003; 9(2): 153-63.
[http://dx.doi.org/10.1179/oeh.2003.9.2.153] [PMID: 12848244]

[229] Crossgrove J, Zheng W. Manganese toxicity upon overexposure. NMR Biomed 2004; 17(8): 544-53.
[http://dx.doi.org/10.1002/nbm.931] [PMID: 15617053]

[230] Finley JW, Davis CD. Manganese deficiency and toxicity: are high or low dietary amounts of manganese cause for concern? Biofactors 1999; 10(1): 15-24.
[http://dx.doi.org/10.1002/biof.5520100102] [PMID: 10475586]

[231] Friedman BJ, Freeland-Graves JH, Bales CW, *et al.* Manganese balance and clinical observations in young men fed a manganese-deficient diet. J Nutr 1987; 117(1): 133-43.
[http://dx.doi.org/10.1093/jn/117.1.133] [PMID: 3819860]

[232] Penland JG, Johnson PE. Dietary calcium and manganese effects on menstrual cycle symptoms. Am J Obstet Gynecol 1993; 168(5): 1417-23.
[http://dx.doi.org/10.1016/S0002-9378(11)90775-3] [PMID: 8498421]

[233] Sloot WN, Gramsbergen JB. Axonal transport of manganese and its relevance to selective neurotoxicity in the rat basal ganglia. Brain Res 1994; 657(1-2): 124-32.
[http://dx.doi.org/10.1016/0006-8993(94)90959-8] [PMID: 7820609]

[234] Sikk K, Haldre S, Aquilonius SM, Taba P. Manganese-Induced Parkinsonism due to Ephedrone Abuse. Parkinsons Dis 2011; 2011: 865319.
[http://dx.doi.org/10.4061/2011/865319] [PMID: 21403909]

[235] Lucchini RG, Dorman DC, Elder A, Veronesi B. Neurological impacts from inhalation of pollutants and the nose-brain connection. Neurotoxicology 2012; 33(4): 838-41.
[http://dx.doi.org/10.1016/j.neuro.2011.12.001] [PMID: 22178536]

[236] Zoni S, Bonetti G, Lucchini R. Olfactory functions at the intersection between environmental exposure to manganese and Parkinsonism. J Trace Elem Med Biol 2012; 26(2-3): 179-82.
[http://dx.doi.org/10.1016/j.jtemb.2012.04.023] [PMID: 22664337]

[237] Au C, Benedetto A, Aschner M. Manganese transport in eukaryotes: the role of DMT1. Neurotoxicology 2008; 29(4): 569-76.
[http://dx.doi.org/10.1016/j.neuro.2008.04.022] [PMID: 18565586]

[238] Gavin CE, Gunter KK, Gunter TE. Manganese and calcium efflux kinetics in brain mitochondria. Relevance to manganese toxicity. Biochem J 1990; 266(2): 329-34.
[http://dx.doi.org/10.1042/bj2660329] [PMID: 2317189]

[239] Gavin CE, Gunter KK, Gunter TE. Mn2+ sequestration by mitochondria and inhibition of oxidative phosphorylation. Toxicol Appl Pharmacol 1992; 115(1): 1-5.
[http://dx.doi.org/10.1016/0041-008X(92)90360-5] [PMID: 1631887]

[240] Chen JY, Tsao GC, Zhao Q, Zheng W. Differential cytotoxicity of Mn(II) and Mn(III): special reference to mitochondrial [Fe-S] containing enzymes. Toxicol Appl Pharmacol 2001; 175(2): 160-8.
[http://dx.doi.org/10.1006/taap.2001.9245] [PMID: 11543648]

[241] Gunter TE, Gerstner B, Lester T, *et al.* An analysis of the effects of Mn2+ on oxidative phosphorylation in liver, brain, and heart mitochondria using state 3 oxidation rate assays. Toxicol Appl Pharmacol 2010; 249(1): 65-75.
[http://dx.doi.org/10.1016/j.taap.2010.08.018] [PMID: 20800605]

[242] Milatovic D, Zaja-Milatovic S, Gupta RC, Yu Y, Aschner M. Oxidative damage and

neurodegeneration in manganese-induced neurotoxicity. Toxicol Appl Pharmacol 2009; 240(2): 219-25.
[http://dx.doi.org/10.1016/j.taap.2009.07.004] [PMID: 19607852]

[243]   Burton NC, Guilarte TR. Manganese neurotoxicity: lessons learned from longitudinal studies in nonhuman primates. Environ Health Perspect 2009; 117(3): 325-32.
[http://dx.doi.org/10.1289/ehp.0800035] [PMID: 19337503]

[244]   Burton NC, Schneider JS, Syversen T, Guilarte TR. Effects of chronic manganese exposure on glutamatergic and GABAergic neurotransmitter markers in the nonhuman primate brain. Toxicol Sci 2009; 111(1): 131-9.
[http://dx.doi.org/10.1093/toxsci/kfp124] [PMID: 19520674]

[245]   Paris I, Segura-Aguilar J. The role of metal ions in dopaminergic neuron degeneration in Parkinsonism and Parkinson's disease. Chem 2011; 142: 365.

[246]   Criswell SR, Perlmutter JS, Videen TO, *et al.* Reduced uptake of [^{18}F]FDOPA PET in asymptomatic welders with occupational manganese exposure. Neurology 2011; 76(15): 1296-301.
[http://dx.doi.org/10.1212/WNL.0b013e3182152830] [PMID: 21471467]

[247]   Kern CH, Stanwood GD, Smith DR. Preweaning manganese exposure causes hyperactivity, disinhibition, and spatial learning and memory deficits associated with altered dopamine receptor and transporter levels. Synapse 2010; 64(5): 363-78.
[http://dx.doi.org/10.1002/syn.20736] [PMID: 20029834]

[248]   Harischandra DS, Ghaisas S, Zenitsky G, *et al.* manganese-induced neurotoxicity: new insights into the triad of protein misfolding, mitochondrial impairment, and neuroinflammation. Front Neurosci 2019; 13: 654.
[http://dx.doi.org/10.3389/fnins.2019.00654] [PMID: 31293375]

[249]   Rosenstock HA, Simons DG, Meyer JS. Chronic manganism. Neurologic and laboratory studies during treatment with levodopa. JAMA 1971; 217(10): 1354-8.
[http://dx.doi.org/10.1001/jama.1971.03190100038007] [PMID: 4998860]

[250]   Hine CH, Pasi A. Manganese intoxication. West J Med 1975; 123(2): 101-7.
[PMID: 1179714]

[251]   Banta RG, Markesbery WR. Elevated manganese levels associated with dementia and extrapyramidal signs. Neurology 1977; 27(3): 213-6.
[http://dx.doi.org/10.1212/WNL.27.3.213] [PMID: 557755]

[252]   Klos KJ, Chandler M, Kumar N, Ahlskog JE, Josephs KA. Neuropsychological profiles of manganese neurotoxicity. Eur J Neurol 2006; 13(10): 1139-41.
[http://dx.doi.org/10.1111/j.1468-1331.2006.01407.x] [PMID: 16987168]

[253]   Guilarte TR. Manganese and Parkinson's disease: a critical review and new findings. Environ Health Perspect 2010; 118(8): 1071-80.
[http://dx.doi.org/10.1289/ehp.0901748] [PMID: 20403794]

[254]   da Silva CJ, da Rocha AJ, Mendes MF, Braga AP, Jeronymo S. Brain manganese deposition depicted by magnetic resonance imaging in a welder. Arch Neurol 2008; 65(7): 983.
[PMID: 18625873]

[255]   de Bie RM, Gladstone RM, Strafella AP, Ko JH, Lang AE. Manganese-induced Parkinsonism associated with methcathinone (Ephedrone) abuse. Arch Neurol 2007; 64(6): 886-9.
[http://dx.doi.org/10.1001/archneur.64.6.886] [PMID: 17562938]

[256]   Colosimo C, Guidi M. Parkinsonism due to ephedrone neurotoxicity: a case report. Eur J Neurol 2009; 16(6): e114-5.
[http://dx.doi.org/10.1111/j.1468-1331.2009.02606.x] [PMID: 19453409]

[257]   Meral H, Kutukcu Y, Atmaca B, Ozer F, Hamamcioglu K. Parkinsonism caused by chronic usage of intravenous potassium permanganate. Neurologist 2007; 13(2): 92-4.

[http://dx.doi.org/10.1097/01.nrl.0000253089.20746.a8] [PMID: 17351530]

[258] Sanotsky Y, Lesyk R, Fedoryshyn L, Komnatska I, Matviyenko Y, Fahn S. Manganic encephalopathy due to "ephedrone" abuse. Mov Disord 2007; 22(9): 1337-43.
       [http://dx.doi.org/10.1002/mds.21378] [PMID: 17566121]

[259] Selikhova M, Fedoryshyn L, Matviyenko Y, *et al.* Parkinsonism and dystonia caused by the illicit use of ephedrone--a longitudinal study. Mov Disord 2008; 23(15): 2224-31.
       [http://dx.doi.org/10.1002/mds.22290] [PMID: 18785245]

[260] Stepens A, Logina I, Liguts V, *et al.* A Parkinsonian syndrome in methcathinone users and the role of manganese. N Engl J Med 2008; 358(10): 1009-17.
       [http://dx.doi.org/10.1056/NEJMoa072488] [PMID: 18322282]

[261] Sikk K, Taba P, Haldre S, *et al.* Clinical, neuroimaging and neurophysiological features in addicts with manganese-ephedrone exposure. Acta Neurol Scand 2010; 121(4): 237-43.
       [http://dx.doi.org/10.1111/j.1600-0404.2009.01189.x] [PMID: 20028341]

[262] Yildirim EA, Eşsizoğlu A, Köksal A, Doğu B, Baybaş S, Gökalp P. Turk Psikiyatr Derg 2009; 20(3): 294-8. [Chronic manganese intoxication due to methcathinone (ephedron) abuse: a case report].

[263] Stepens A, Stagg CJ, Platkajis A, Boudrias MH, Johansen-Berg H, Donaghy M. White matter abnormalities in methcathinone abusers with an extrapyramidal syndrome. Brain 2010; 133(Pt 12): 3676-84.
       [http://dx.doi.org/10.1093/brain/awq281] [PMID: 21036949]

[264] Iqbal M, Monaghan T, Redmond J. Manganese toxicity with ephedrone abuse manifesting as parkinsonism: a case report. J Med Case Reports 2012; 6: 52.
       [http://dx.doi.org/10.1186/1752-1947-6-52] [PMID: 22313512]

[265] Wong BS, Chen SG, Colucci M, *et al.* Aberrant metal binding by prion protein in human prion disease. J Neurochem 2001; 78(6): 1400-8.
       [http://dx.doi.org/10.1046/j.1471-4159.2001.00522.x] [PMID: 11579148]

[266] Bouchard MF, Sauvé S, Barbeau B, *et al.* Intellectual impairment in school-age children exposed to manganese from drinking water. Environ Health Perspect 2011; 119(1): 138-43.
       [http://dx.doi.org/10.1289/ehp.1002321] [PMID: 20855239]

[267] Khan K, Factor-Litvak P, Wasserman GA, *et al.* Manganese exposure from drinking water and children's classroom behavior in Bangladesh. Environ Health Perspect 2011; 119(10): 1501-6.
       [http://dx.doi.org/10.1289/ehp.1003397] [PMID: 21493178]

[268] Menezes-Filho JA, Novaes CdeO, Moreira JC, Sarcinelli PN, Mergler D. Elevated manganese and cognitive performance in school-aged children and their mothers. Environ Res 2011; 111(1): 156-63.
       [http://dx.doi.org/10.1016/j.envres.2010.09.006] [PMID: 20943219]

[269] Claus Henn B, Ettinger AS, Schwartz J, *et al.* Early postnatal blood manganese levels and children's neurodevelopment. Epidemiology 2010; 21(4): 433-9.
       [http://dx.doi.org/10.1097/EDE.0b013e3181df8e52] [PMID: 20549838]

[270] Bertini I, Sigel A, Sigel H, Eds. Handbook on metalloproteins Marcel Dekker. New York, Basel: Springer 2001; pp. 836-7.
       [http://dx.doi.org/10.1201/9781482270822]

[271] Barceloux DG. Nickel. J Toxicol Clin Toxicol 1999; 37(2): 239-58.
       [http://dx.doi.org/10.1081/CLT-100102423] [PMID: 10382559]

[272] Muñoz A, Costa M. Elucidating the mechanisms of nickel compound uptake: a review of particulate and nano-nickel endocytosis and toxicity. Toxicol Appl Pharmacol 2012; 260(1): 1-16.
       [http://dx.doi.org/10.1016/j.taap.2011.12.014] [PMID: 22206756]

[273] Denkhaus E, Salnikow K. Nickel essentiality, toxicity, and carcinogenicity. Crit Rev Oncol Hematol 2002; 42(1): 35-56.

[http://dx.doi.org/10.1016/S1040-8428(01)00214-1] [PMID: 11923067]

[274] Samal L, Mishra C. Significance of nickel in livestock health and production. Inter J Agro Vet Med Sci 2011; 5: 349-61.
[http://dx.doi.org/10.5455/ijavms.20110331111304]

[275] Beltrán González AN, López Pazos MI, Calvo DJ. Reactive Oxygen Species in the Regulation of the GABA Mediated Inhibitory Neurotransmission. Neuroscience 2020; 439: 137-45.
[http://dx.doi.org/10.1016/j.neuroscience.2019.05.064] [PMID: 31200105]

[276] Xu SC, He MD, Zhong M, *et al.* Melatonin protects against Nickel-induced neurotoxicity *in vitro* by reducing oxidative stress and maintaining mitochondrial function. J Pineal Res 2010; 49(1): 86-94.
[http://dx.doi.org/10.1111/j.1600-079X.2010.00770.x] [PMID: 20536687]

[277] He MD, Xu SC, Zhang X, *et al.* Disturbance of aerobic metabolism accompanies neurobehavioral changes induced by nickel in mice. Neurotoxicology 2013; 38: 9-16.
[http://dx.doi.org/10.1016/j.neuro.2013.05.011] [PMID: 23727075]

[278] Kahloula K, Adli DEH, Slimani M, Terras H, Achour S. Effect of chronic exposure to nickel on the neurobehavioralfunctions in Wistar rats during the development period. Toxicol Anal et Clin 2014; 26(4): 186-92.

[279] Cragle DL, Hollis DR, Newport TH, Shy CM. A retrospective cohort mortality study among workers occupationally exposed to metallic nickel powder at the Oak Ridge Gaseous Diffusion Plant. IARC Sci Publ 1984; (53): 57-63.
[PMID: 6532993]

[280] Forte G, Alimonti A, Pino A, *et al.* Metals and oxidative stress in patients with Parkinson's disease. Ann Ist Super Sanita 2005; 41(2): 189-95.
[PMID: 16244392]

[281] Maass F, Michalke B, Leha A, *et al.* Elemental fingerprint as a cerebrospinal fluid biomarker for the diagnosis of Parkinson's disease. J Neurochem 2018; 145(4): 342-51.
[http://dx.doi.org/10.1111/jnc.14316] [PMID: 29388213]

[282] Gupta V, Ansari NG, Garg RK, Khattri S. Determination of Cd, Cr, Pb and Ni contents among Parkinson's disease individuals: a case-control study. Int J Neurosci 2017; 127(9): 770-5.
[http://dx.doi.org/10.1080/00207454.2016.1251917] [PMID: 27819176]

CHAPTER 6

Developmental Toxicity of Aluminium and other Metals: Areas Unexplored

Laraib Liaquat[1,2], Zehra Batool[3,*] and Saida Haider[1]

[1] *Neurochemistry and Biochemical Neuropharmacology Research Unit, Department of Biochemistry, University of Karachi, Karachi, Pakistan*

[2] *Multidisciplinary Research Lab, Bahria University Medical and Dental College, Bahria University, Karachi, Pakistan*

[3] *Dr. Panjwani Center for Molecular Medicine and Drug Research, International Center for Chemical and Biological Sciences, University of Karachi, Karachi, Pakistan*

Abstract: Reproduction and developmental damage has irreversible consequences compared to other body functions and may have adverse effects throughout life. In some circumstances, the damage passes from generation to generation. Many environmental agents contribute to developmental toxicities such as toxic metals, insecticides or pesticides, commercial or industrial pollutants, and air pollutants. Increased urbanization and industrialization have led to the accumulation of toxic metals in the environment. Widespread use of heavy metals in different fields such as agriculture, domestic, medical, industrial, and technological applications have resulted in increased exposure of heavy metals to the human population. Environmental exposure to heavy metals is extensively linked to toxic effects on mammalian embryos. Metals such as lead, cadmium, mercury and arsenic are known developmental toxicants that intensely affect fetal and embryonic development and cause certain malformations in developing embryo even at low concentrations. Other metals such as uranium, cobalt, lithium, Aluminium, manganese, and copper are also reported to induce developmental consequences, including neurobehavioral abnormalities, neural tube defects, fetal growth retardation, skeletal deformation, preterm or delayed birth, and still birth or postnatal death. Heavy metal developmental toxicity depends on different factors, including dose, duration, and route of exposure. Hence, heavy metals are known to be toxic to fetal and embryonic tissues and can produce serious teratogenicity in mammals; however, not much attention has been given to this topic. This chapter, therefore, summarizes the developmental toxicity of heavy metals on the mammalian system and their teratogenic mechanism in growing embryos.

Keywords: Developmental Toxicity, Heavy Metals, Toxicity Mechanism.

* **Corresponding author Zehra Batool**: Dr. Panjwani Center for Molecular Medicine and Drug Research, International Center for Chemical and Biological Sciences, University of Karachi, Karachi, Pakistan; Tel:+92-21-99261737; E-mail: xehra_batool@yahoo.com

INTRODUCTION

Developmental toxicity is the science to understand the effects of environmental insults to interfere with the normal developmental process which may also result in adverse effects in the next generations. The genetic, nutritional, infectious, and chemical factors are thought to cause congenital abnormalities in humans. Such manifestations of developmental abnormalities have also been observed in animals. Exposure to heavy metals by different environmental factors particularly during the gestational process can induce deleterious and sometimes irreversible damages to the developing embryo. These effects may also result in further developmental abnormalities in the postnatal stage.

The conceptus, which is defined as the embryo and embryo-derived embryonic tissues, is highly susceptible to the toxic effects of heavy metals. These metals can cross the placenta and can directly interact with conceptus. The ultimate result of abnormal development may result in functional disorder, growth retardation, malformation, or death. Therefore, it is important to identify possible developmental toxicities of heavy metals that are commonly present in our environment. Most of the studies regarding metal toxicity have focused on the brain, hepatic, or kidney functions. However, little is known about the developmental toxicity induced by metals. Teratogenic agents act through a particular mechanism on developing cells and tissues to initiate embryogenic aberrations. The understanding of metal-induced defects in the placenta and fetal development is important to develop preventive and control approaches to ensure normal embryogenesis. In the following sections, teratogenic properties of those heavy metals have been explored with which the general population comes in contact almost on a daily basis.

ALUMINIUM

Aluminium is the third most abundant metal and constitutes about 8% of the earth's crust [1]. According to the World Health Organization's report, the living body gets exposed to Aluminium through food, antacids, cooking utensil, and deodorants besides occupational exposure such as defense-related factories, guns, and automobiles [2]. Aluminium compounds are also used in water purification processes that lead to increased Aluminium levels in drinking water. Aluminium compounds can reach systematic circulation *via* different routes such as dermal absorption, ingestion, and intramuscular injection [1].

Teratogenic Nature of Aluminium

Aluminium has the potential to cross the placental barrier and accumulate in fetal tissues [3]. Environmental and dietary exposure to Aluminium during the maternal period causes developmental toxicities in mammals. Aluminium causes severe developmental syndromes, including neurodevelopment disturbances such as mental retardation, skeletal and soft tissue abnormalities, and growth retardation [4]. Oral exposure of a large amount of Aluminium to pregnant women is of special concern. Antacids are normally prescribed to pregnant women to treat dyspepsia and associated dyspeptic symptoms. Aluminium-containing antacids are associated with increased accumulation of Aluminium in maternal blood. Overexposure to Aluminium during pregnancy through antacids is linked with embryonic and fetal toxicities [5]. Therefore, due to toxicities associated with Aluminium exposure, it is generally suggested to limit the intake of Aluminium-containing antacids during pregnancy [6].

Aluminium exposure during the gestational period to female rats induces prenatal, teratogenic and postnatal adverse effects [7]. Studies have reported that Aluminium exposure can induce fetal abnormalities, growth retardation, delay in ossification in rats and mice [8]. Reported malformations are attributed to the transplacental passage of Aluminium [9]. Aluminium nitrate in mice induces abortions, preterm delivery, and fetal death. Aluminium-related developmental toxicities in animals depend on the route of administration and nature of Aluminium compound administered [6]. Oral administration of Aluminium chloride to female rats during organogenesis, fetal, and lactation period resulted in post implantation deaths, resorptions, morphological alterations along with the visceral and skeletal anomalies. Delayed birth, dystocia, neurobehavioral and respiratory disturbances are also observed following Aluminium exposure in animals during fetal development [7].

Mechanism of Teratogenic Activity

A number of possible mechanisms have been suggested regarding the teratogenic nature of Aluminium. Experimental work on Aluminium treated pregnant animals reported that Aluminium oral exposure during pregnancy alters tissue distribution of essential trace elements. Higher concentration of copper was found in the brain, whereas calcium, magnesium, manganese concentrations were significantly high in renal tissue. Such changes ultimately produce negative outcomes on fetal metabolism [10]. Aluminium competes with essential cations such as iron, calcium, and magnesium to bind with fluoride and phosphate anions. Such interactions alter the biological active mechanism of essential ions including uptake, distribution, and excretion [11]. Essential ions are required for proper bone formation; however, interactions of Aluminium with cations and anions produce fetal bone malformation [11]. It is also associated with inhibition of the parathyroid gland and osteoblast formation [12]. Aluminium accumulation in the fetal brain has negative consequences on neuronal cell nuclei where it interacts with phosphate group in nucleic acid and phosphorylated proteins. It is reported that this interaction results in condensation of brain chromatin configurations, inhibits protein synthesis, and reduces transcription in nerve cells, and leads to reduced generation of nerve cells from progenitor cells as well as reduces division of glial and epithelial cells, which is important for normal brain functions and development (Fig. **1**). Therefore, high accumulation of Aluminium in fetal brain inhibits brain growth and development, and produces neurological alterations in offspring [7].

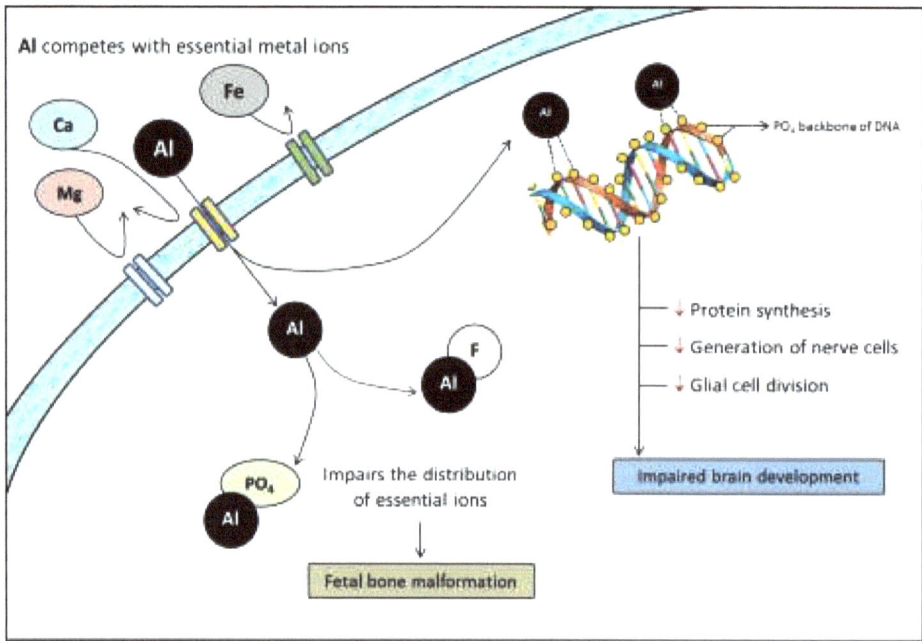

Fig. (1). Aluminium passes through the placental barrier and easily enters into the fetal cells by competing with essential metal ions. This disturbs the balance of metals required for normal bone development. Aluminium also affects the gene expression of vital proteins necessary for normal brain development.

ARSENIC

Inorganic arsenic is more toxic than organic and both trivalent and pentavalent species of inorganic arsenic are teratogenic and cause serious developmental toxicities [13]. Human body gets exposed to arsenic *via* different routes including food, water, medicinal and occupational exposure. Environmental arsenic exposure creates a teratogenic threat to the human population. Contaminated groundwater is the main and foremost important source of arsenic exposure in Asian countries to the human population [14].

Teratogenic Nature of Arsenic

Human and experimental studies have reported the ability of arsenic to pass through the placenta at concentrations exceeding that of the maternal blood [15]. Exposure at acute high-dose and chronic low-dose during pregnancy has been associated with pre and post-natal mortality [16]. Pregnancy loss, developmental impairment, and low birth weight have been reported following chronic low-dose arsenic exposure [17]. Congenital heart malformations and neural tube defects are the most severe developmental impairment reported in offspring following maternal arsenic exposure through drinking water. Arsenic exposure in prenatal and early childhood may produce severe neurodevelopment abnormalities at young age [18 - 20]. Neural tube defects are the most common birth defects worldwide. They are characterized by high rates of mortality and severe disabilities in those who survived [14]. Exposure during the neural tube closure may also result in skeletal disruption mainly in axial components [21].

Studies on laboratory animals have extensively reported the potent teratogenic nature of arsenic [13]. Arsenic can cross placenta and accumulates in the neuroepithelium of embryos. It causes developmental toxicities in neural plate by producing reactive oxygen species and disruption of glucose metabolism that ultimately results in programmed cell death and failure of neural tube closure [22]. Studies have reported developmental malformation in four species including hamsters, rabbits, rat, and mice. Malformation pattern depends on arsenic dose, route, and day of exposure during gestation and characteristic outcomes are death and growth retardation [23]. Gefrides *et al* [24] have reported neural tube defects due to teratogenic effects of arsenic exposure in Splotch embryos.

Mechanism of Teratogenic Activity

Various genetic and biological pathways are associated with the teratogenic mechanism of arsenic and responsible for multifaceted birth defects [25]. Arsenic interferes with phosphorylation reactions by acting as a phosphate analog, interacts with enzymatic sulfhydryl groups and produces inhibitory effect on various cellular processes. Both species of inorganic arsenic are extremely mitochondriotoxic and disrupt mitochondrial functions and produces oxidative stress which in turn is associated with macromolecular damage and signaling disregulation [26].

LEAD

Lead is a non-essential heavy metal and its environmental pollution arises from many sources and results in multiple health and environmental problems [27]. Lead is used in petroleum industries for petrol (gasoline) processing, paint industry and emission from automobiles fuel results in increased contamination of lead in the environment. Lead gets access to the living body through inhalation and by food and drinks *via* gastrointestinal tract [28]. It causes severe developmental toxicities such as developmental delay and major neurological disorders in mammals and amphibians [29].

Teratogenic Nature of Lead

Lead toxicity in pregnant women is a major public health issue and especially due to the large immigrant population in urban setting it is becoming a serious health concern [30]. Experimental and clinical studies have reported prenatal lead exposure as a causative factor in the incidence of brain defects [31, 32]. Various cohort and case-control studies have indicated the link of prenatal lead exposure and neural tube defects in offspring. A Norwegian cohort study has reported that the frequency of neural tube defects in offspring was more in women occupationally exposed to lead as compared to non-exposed women [33]. Case-control study in United Kingdom has found that the risk of pregnancies with neural tube defects specially anencephaly was more in women living in electoral wards with a higher proportion of lead in drinking water (<10mg/L lead) compared to control [34]. Another case control study in Turkey on women with terminated pregnancies in second trimester after ultrasonographic diagnosis of neural tube defect in the growing fetus has reported that these women also had significantly higher levels of blood lead compared to women with normal pregnancy. So there must be a close link between high levels of lead in blood with the incidence of neural tube defects [35]. Maternal whole blood or plasma lead levels in the first trimester of pregnancy can also be used as a predictor of neurodevelopment status as high fetal lead exposure has severe negative effects on neurodevelopmental process [36].

Increased maternal lead exposure is associated with infertility, high risk of abortion, reduced fetal growth, low birth weight, preterm delivery, and disruption in neurobehavioral development. Studies have suggested a link between paternal lead exposure and congenital malformation [37]. Lead has negative consequences on fertility as it reduces fertility in both males and females. Blood level exceeding 40 µg/dl of lead is responsible for fertility issues in males and blood level exceeding 30 µg/dl of lead in women is responsible for spontaneous abortion [38]. Maternal lead exposure 5-9µg/dl is associated with a higher risk of reproductive issues. High level of lead in maternal blood is also linked with complicated pregnancies because high blood level leads to a higher concentration in trophoblastic placental tissue that ultimately results in premature rupture of membrane and preterm delivery [39] whereas, lead exposure during the period of spermatogenesis is associated with a congenital malformation in offspring. Other malformations related to lead exposure include congenital heart disease, oral cleft, club foot, and polydactyly [40]. Increased risk of male cleft palate is also reported following paternal lead exposure. Long running perspective studies of child development and prenatal lead exposure have associated prenatal lead exposure with delayed neurodevelopment in offspring. Moreover, an inverse relation of pre- and post-natal blood levels with neurobehavioral status in offspring has also been reported [37].

Lead has ability to cross the placenta and can cause severe developmental toxicities, as evident from animal studies [27]. Experimental exposure of lead to Estuarine crabs *(Chasmagnathus granulates)* at a low level (0.01-1 mg/L) during early, late or whole embryonic development results in several morphological abnormalities in hatched larvae and the teratogenic effects of lead were more prominent in early period of embryogenesis [24]. The morphological changes including eye malformation, reduced pigmentation, irregular head shape, and notochordal defects have also been observed in African catfish exposed to lead at the dose of 0.3 and 0.5 mg/L [41]. Studies have indicated that lead can interact with luteinizing hormone, and can impair its function. Experimental work on time-impregnated female SD rats by administering lead acetate in drinking water has shown that lead acetate can produce adverse effects on gonadal function resulting in delayed sexual maturity, low plasma testosterone and estradiol levels, and disturbed estrus cycle during puberty in offspring [42]. Lead exposure also has adverse effects on male rats including low sperm production, reduced testicular spermatids, altered prostatic functions, and alteration in serum testosterone levels [28]. Oral supplementation of lead acetate to mice results in delayed growth and maturation of ovarian follicles [43].

Mechanism of Teratogenic Activity

Lead alters the basic mechanism of protein metabolism and lowers the seminal plasma protein, indicated by rise in free amino acid in plasma. This eventually leads to disturbance in cellular nutritional mechanism which is required for survival of sperm cells and their functions [44]. The toxicity induced by lead is attributed to its ability to mimic calcium and associated signaling pathway. Through calcium channel, lead enters into the cell and binds with calcium-dependent protein kinases such as protein kinase C. This binding alters the functioning of kinase enzymes and thus interrupts physiological functions [45]. Lead has shown to affect the hypothalamic-pituitary-gonadal axis by altering the signaling between hypothalamus and pituitary thereby modulating the release of gonadotropin-releasing hormone in male rats and affecting spermatogenesis [46]. The mechanism for detrimental effects of lead on growing embryo is postulated by the interaction of lead with human chorionic gonadotropin (HCG) in placenta.

HCG hormone is primarily required for the maintenance of pregnancy by producing progesterone from corpus luteum. Lead has shown to change the secondary structure of HCG leading to the loss of functional integrity of HCG and pregnancy [47]. Oxidative stress is also implicated in the lead-induced toxicity. Short-term exposure to lead has been shown to increase the activity of catalase and superoxide dismutase whereas, repeated exposure results in inhibition of antioxidant enzymes and reduced glutathione. Long-term exposure to lead disrupts the balance of prooxidants *versus* antioxidants and mediates damage to vital cellular components leading to cell death [45].

MERCURY

Mercury (Hg) is a highly toxic metal generally recognized as silver liquid at room temperature and mainly present in thermometer, manometers, barometers, fluorescent light bulb, and dental amalgam. It enters the body through inhalation in the form of vapors, as a primary route to get access to the human body. Occupational exposure, fossil fuel combustion, gold refining, and mercury mining are also regarded as sources of mercury exposure. It is also used in traditional medicine, antiseptics, skin ointments, diuretics, and in some vaccination as a preservative. However, the general population gets exposed to mercury in the form of vapors released during dental amalgam [48] and fish consumption [49].

Teratogenic Nature of Mercury

Mercury is highly soluble in lipid so it can easily penetrate through placental barrier and can further pass to fetal tissues. Inside the fetal tissue, mercury undergoes oxidation and converts to Hg^{+2}, oxidized form of mercury, which can easily accumulate in fetus and can also re-cross placental barrier. The accumulation of mercury is more in fetus than mother and exposure of mercury vapors to fetus can induce adverse developmental effects in the fetus [50]. Mercury can cause multiple organ toxicity. Studies have reported that mercuric compounds can induce severe hepatotoxic, nephrotoxic, and neurotoxic effects [48]. Fish is the main source of methyl mercury and studies on human exposure have reported that methyl mercury can induce severe developmental toxicities in humans and can result in methyl mercury poisoning [49, 50]. Methyl mercury is recognized as a teratogenic agent because it can cause neurological disturbance and developmental abnormalities in infants [51]. Researchers have noted that the fetal brain is more susceptible to mercury induced-damage compared to the adult brain. It disturbs the division and migration of neuronal cells during brain development and alters the cytoarchitecture of fetal brain [52]. The research report on prenatal methyl mercuric exposure through fish consumption showed that it produces neuropsychological changes and because of such damaging effects, therefore, Food and Drug Administration has recommended pregnant women, nursing mothers, and young children to avoid fish consumption with high mercury content such as whale, shark, and tilefish [53]. Toxicity caused by ethyl mercury is another concern in the medical world. Infants and children get exposed to ethyl mercury in the form of thimerosal through a vaccine that produces brain defects [49].

Mercury is highly teratogenic in any form and the teratogenic effects induced by mercury include stillbirth, spontaneous abortion, and congenital malformation as evident from animal studies [54]. Mercury exposure to pregnant rats leads to increased incidence of resorptions and results in reduced birth weight and litter size. It causes an alteration in placental functions that ultimately affect growing fetus and may produce severe adverse pregnancy outcomes [50]. Baranski and Szymczak observed that prenatal exposure of mercury vapors (Hg^{o}) for 21 days induced severe toxicities and rat pups died within 6 days after birth [55]. Observed effects were due to high levels of mercury in maternal blood that further accumulate in fetus during developmental period. Experimental work on Medaka fish has further reported that both organic (MeHg) and inorganic ($HgCl_2$) forms of mercury are teratogenic in nature and can induce developmental toxicities including reduced eye pigmentation, failure of swim bladder inflation, elongated heart, and pericardial edema [56]. Other malformations and developmental alterations associated with methyl mercury exposure in animals following intraperitoneal or oral administration include cleft palate, brain lesions, generalized edema, asymmetrical sternebrae, wavy ribs, and decreased ossification of parietal and occipital bones [57].

Mechanism of Teratogenic Activity

Inhaled vapors of mercury can easily transport into the blood through pulmonary circulation and 80% of the inhaled mercury is retained and absorbed into the blood. In the blood some amount of mercury is trapped by the erythrocytes whereas remaining mercury is distributed to other tissues [58]. As mercury is highly lipid soluble so it can easily pass through the cell membrane. Inside the cell, it is acted upon by cystolic catalase enzyme that converts it into mercuric (Hg^{+2}) or oxidized form of mercury. This oxidized form of mercury is highly reactive in nature and can produce reactive oxygen species which in turn disturb and denature many important macromolecules such as protein and enzymes and ultimately affect organ functions [59]. Oxidized mercury can bind with the protein sulfhydryl group. Conceptus or pregnant female produce essential protein that is important for growing fetus but binding or interaction of Hg^{+2} with such protein may alter their function and ultimately affect the developmental process (Fig. **2**).

A high dose of mercury can also interact with the growing fetus non-specifically and may result in resorption [50].

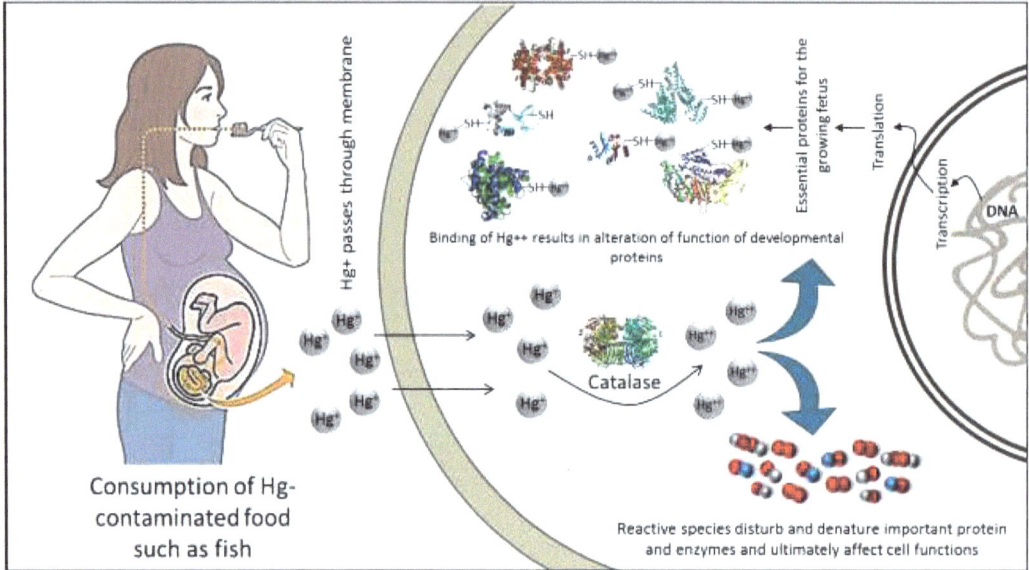

Fig. (2). Illustration of mechanism of teratogenic activity of mercury.

CADMIUM

Cadmium is another heavy metal environmental pollutant that also has the potential to induce teratogenic effects. Due to increased industrialization and modernization of life style, a common person is exposed to cadmium toxicity almost on the daily basis. The common source of cadmium exposure includes the ingestion of contaminated food and water. The entire food chain is exposed to cadmium toxicity due to unprocessed spillage of industrial waste in water streams. Tobacco smoke is also considered as the major source of cadmium exposure in active as well as in passive smokers. The toxicity of cadmium accounts to the immediate penetration into the body organs after its exposure and its long biological half-life ranging from 75 days to 26 years.

Teratogenic Nature of Cadmium

Cadmium has been repeatedly demonstrated as a toxicant for fetal development. The cadmium can penetrate into the placenta and can induce embryopathogenesis. The content of cadmium has been found to be more in women than in men. Studies have demonstrated that it can affect different stages of embryo development from pre-implantation to organogenesis.

Food is the major source of cadmium exposure during gestational period in women. It has been shown in a survey carried out by Moynihan and co-workers that the blood content of cadmium in pregnant women was directly correlated with the dietary concentration of cadmium which was also observed in serum of offspring [60]. Smoking is also one of the main sources of cadmium exposure. Placentas of 56 smoker and non-smoker women were analyzed. It was found that the placenta of smoker mothers exhibited high concentration of lead, copper, iron, and cadmium. The heavy metal concentration was inversely correlated with progesterone levels representing the possible reason for recurrent abortion [61, 62]. The complications in pregnancy have also found to be positively correlated with the blood cadmium content. Sukhodolska reported 2.8 times increased cadmium concentration in women with complicated pregnancy as compared uncomplicated cases whereas, women with threatened abortion exhibited 3.2 times higher cadmium concentration than the normal pregnancy cases [63]. Cadmium crosses the placental barrier and accumulates in placenta resulting in the impairment of fetal development [64]. The cadmium toxicity not only affects child-bearing women but also induces long-term impairments in offspring. Cadmium can accumulate in infants through breastfeeding leading to systemic diseases. In these offspring the ability of learning and memory is mainly affected due to cadmium exposure [65, 66]. The toxicity of cadmium has also been shown to affect male reproductive system. Sixty infertile males were examined by Akinloye and colleagues and they found a significantly higher concentration of plasma and seminal cadmium levels than the normal fertile male group. The cadmium concentration in infertile males was also negatively correlated with the quality of semen and sperm [67]. Cadmium has been considered as one of the main causes of prostate cancer [68].

The teratogenic property of cadmium has also been consistently reported in various pre-clinical studies. It has been used in embryopathological studies to develop congenital aberrations. Developmental abnormalities have been observed in rodent model at different stages of morula formation. Exposure to cadmium during post-implantation stage of embryo has shown deficits in organogenesis. Craniofacial and ocular deficits, growth abnormalities, notochord retardation, gut and cardiovascular deformities, hypopigmentation, and skeletal irregularities have been observed in various animal models [69, 70]. Defects in the closure of neural tube have also been reported following cadmium exposure in rodents during the gastrulation phase [70, 71]. Cadmium toxicity has also been shown to induce abnormalities in male reproductive system. El-Demerdash and co-workers intoxicated the male rats by cadmium administration and at the end of treatment they observed reduced weight of testes, epididymis, and decreased concentration of sperms which were mostly dead and abnormal [72]. Cadmium as the cause of prostate cancer has also been reported in various preclinical studies. Loss of balance of gonadal hormones and development of testicular necrosis, testicular atrophy, and increased proliferation of interstitial tissue in testes has been observed in a rat model of prostate tumors induced by cadmium exposure [73].

Mechanism of Teratogenic Activity

Human and animal studies have been conducted to elucidate the mechanism of teratogenic activity of cadmium. These studies have revealed that the exposure of cadmium induced changes at multiple cellular levels [74 - 76]. Cadmium exposure has shown to induce epigenetic changes in specific genes that are involved in placental and fetal development. These epigenetic changes are gender dependent and induce different consequences in male and female newborns. The methylation changes of genes in a female fetus result in alterations in organogenesis, morphology, and mineralization, whereas in male fetus these changes can result in alteration in cell-death related genes [77]. Protocadherin, a glycoprotein is abundant in developing embryo and is involved in cell-cell recognition, adhesion, migration, communication, and tissue differentiation. It also plays important role in early embryogenesis and the development of central nervous system. Exposure to cadmium in the early gastrulation process has shown to negatively influence the gene expression of protocadherin. This results in the inhibition of cell proliferation, reduced migration, enhanced apoptosis, and ultimately delays the development of placenta and embryo [78]. Placenta is involved in the development of fetal hypothalamic-pituitary-adrenal axis by regulating the expression of glucocorticoid receptor (NR3C1) and controls cortisol levels in the fetus. Maternal exposure to cadmium is directly associated with the increased methylation of promoter region of NR3C1 and reduces its expression (Fig. **3**). This change in NR3C1 function may result in long-term cognitive and neurobehavioral alterations in offspring [79]. Besides the epigenetic changes of developmental genes in placenta and fetus, alterations of gene expression of various pro-inflammatory cytokines are also involved in the mechanism of cadmium-induced teratogenic activity [80]. The gene expression of TNF-α, interleukin (IL)-6, and IL-8 is upregulated in placenta due to cadmium exposure during gestational process [81 - 83]. Activation of Akt signaling pathway is also associated with cadmium-induced gestational inflammation and impairs placental function and development [84].

Gestational cadmium exposure results in reduced nutrient supply to embryo from placenta leading to delayed embryonic development. Cadmium exposure can reduce nutrient transporters such as glucose transporter (GLUT) and thus reduce the supply of energy and various nutrients required for normal fetal development [85]. Cadmium also disturbs embryonic zinc ion homeostasis. This is achieved by the down-regulation of zinc transporter 1 and 2 in placenta. Zinc is very crucial for the normal development of embryo [86]. Cadmium-induced decreased transportation of zinc leads to improper differentiation of nerves, unusual neuroblasts formation, and abnormal growth and development of embryo [87, 88].

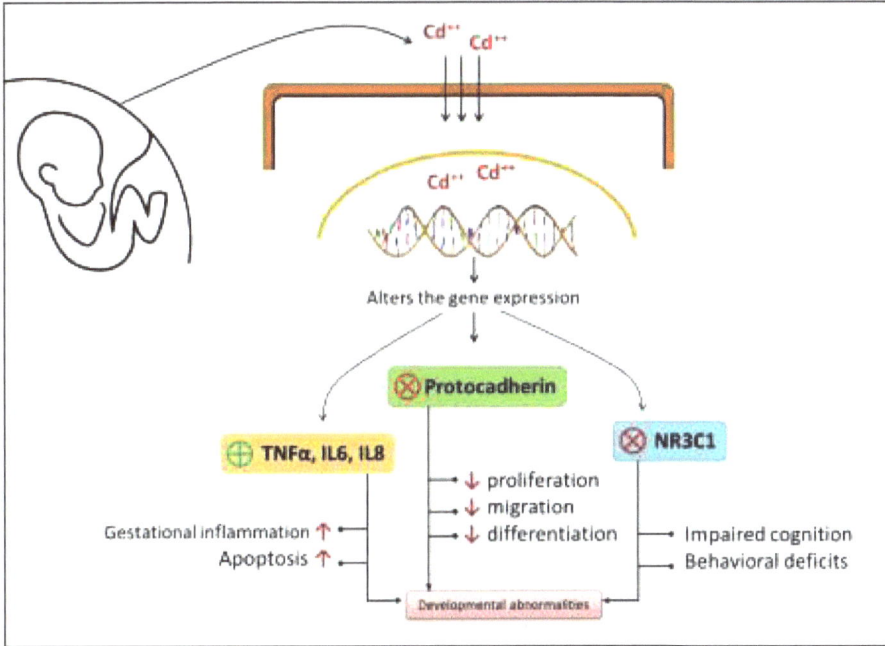

Fig. (3). Illustration of mechanism of teratogenic activity of cadmium ⊗Inhibition; ⊕activation.

URANIUM

Uranium is a heavy metal with radioactive properties. Increased use of uranium in nuclear industries is the main causative factor in occupational exposure to this metal. Nuclear industry has put the general population at a high risk of toxicological effects of uranium either by inhalation in the form of depleted uranium aerosols or through dietary intake [57]. Depleted uranium is a man-made product and is found to be higher in military persons and civilians using munitions [89]. Along with this, it is also used in hospitals as X-ray radiation shield, flaps in commercial aircrafts, in military and non-military airplanes as ballasts where inhalation is the main route of human exposure.

Teratogenic Nature of Uranium

Uranium exists in both soluble and insoluble form, soluble form is associated with chemical toxic effects and insoluble form produces radioactive toxic effects [90]. Due to organotrophic nature, uranium has long term retention ability in target organ mainly kidney and skeletal muscles and can accumulate in various vital organs including the brain and lungs. It has reported genotoxic, carcinogenic, mutagenic, and teratogenic properties [91]. Clinical epidemiological case study by a research team in maternal and children hospital of Basra, Iraq has indicated that exposure of uranium during 1999-2000 in the form of heavy bombardment of uranium munitions resulted in severe malformation in newborns including congenital heart disease, multiple congenital malformations, hydrocephalus, anencephaly, and musculoskeletal disturbances [92]. Another study in Diwaniah, Iraq has reported a large number of births with neural tube defects in the year 2000 [93]. Multiple congenital malformations were also reported among offspring of US veterans of Gulf War. Multiple anomalies were reported including aortic valve stenosis, renal agenesis, or hypoplasia, tricuspid valve insufficiency, hypospadias, and epispadias [94, 95].

Maternal exposure of uranium oxides to female mice through subcutaneous and oral administration cause infertility, fetal toxicity such as congenital malformations including skeletal defects and cleft palate, reduced growth and variations in developmental ossification [96]. Reproductive toxicity is also reported following excessive uranium exposure, such as damage to genetic material, skeletal malformations, and dominant lethality in fetal rats. Studies on male rats reported that uranium has negative effects on testes and germ cells [89].

Mechanism of Teratogenic Activity

The physiological mechanism behind the teratogenic nature of uranium involves its ability to act at molecular level by damaging DNA and RNA at cellular and organ levels, affecting multiple reproductive organs such as testes and placenta [96].

LITHIUM

Lithium is extensively used in air conditioning, chemical and biological laboratories, and in medicines that enhance the probability of lithium getting access to human body. Lithium is classified as class D drug. This category of drug is legal to use during pregnancy but has been associated with severe birth defects. It is among the thirty known teratogenic drugs [97]. Normally, lithium is used in standard treatment of bipolar disorders as mood stabilizing agent [98].

Teratogenic Nature of Lithium

Treatment of pregnant women suffering from bipolar disorder with lithium is a major risk factor for developing fetus [98]. Lithium salts such as lithium carbonate are used in medicine and their long term administration is linked with multiple teratogenic effects [99]. Lithium is the first choice of mood stabilizing agent in pregnant women [100] and prescribed to prevent bipolar disorder during and after pregnancy in postpartum period. Lithium is permeable to the placental barrier and its concentration equilibrates between fetal and maternal blood [101].

Maternal lithium exposure is linked with human fetal risk and physicians have warned its use during pregnancy. The use of lithium in the first trimester of pregnancy is associated with cardiac malformations including Ebstein's anamoly. It is a birth defect characterized by obstruction in the right ventricular outflow tract. A cohort study of 1,325,563 pregnant women has indicated that prenatal exposure of lithium induced cardiac malformations in newborn as compared to un-exposed women. The primary outcome of prenatal lithium exposure is cardiac malformation whereas secondary outcomes are overall major congenital malformations [102]. Two cases per 100 births have been reported as a result of lithium exposure during first trimester of pregnancy. *In utero* exposure of lithium and risk of cardiac malformations is dose dependent, moreover, 900g lithium per day is associated with teratogenic activity [102]. Clinical guidelines discourage breastfeeding in women undergoing lithium treatment [101]. Other birth defects associated with maternal exposure to lithium are hypotonia, poor feeding ability, cyanosis, and respiratory distress syndrome [98]. Risk of preterm birth is also reported in lithium exposure during pregnancy [103]. Exposure of higher dose of lithium in pregnant rats produces craniofacial defects, exencephaly, and developmental abnormalities in blood vessels. Vasculogenesis inhibition and abnormalities of dorsoventral specification are also reported in other vertebrates [104]. Neurodevelopmental deficits have been reported in rats, mice, and zebrafish following prenatal lithium exposure [101].

Mechanism of Teratogenic Activity

Fetal exposure to lithium during the developmental period produces lifetime physiological and metabolic abnormalities [105]. Lithium causes vasculature abnormalities in fetus mainly in first trimester because majority of vasculature formation occurs in the first trimester [106]. Pathophysiological mechanism behind the teratogenic nature of lithium is mainly the involvement of glycogen synthase kinase-3β (GSK3β) enzyme. Lithium inhibits the activity of GSK3β enzyme. GSK3β is an important enzyme of Wnt signaling pathway. Wnt signaling pathway is required for the normal fetal development of cardiac and vascular tissues (Fig. **4**). Any abnormality or inhibition of Wnt pathway produces severe congenital malformations and leads to cardiac abnormalities in the developing fetus [107, 108].

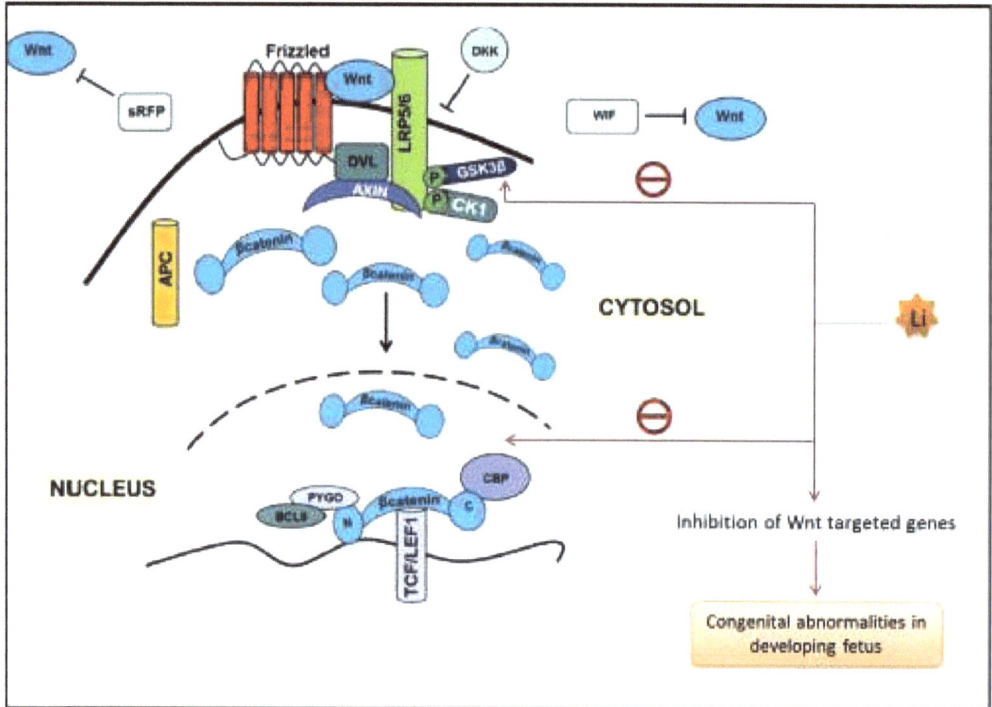

Fig. (4). Wnt signaling pathway regulates crucial aspects of cell fate determination, cell polarity, cell migration, neural patterning and organogenesis during embryonic development. GSK3β is involved in the phosphorylation of downstream signaling proteins. Lithium has the potential to inhibit GSK3β and thus impairs normal fetal development. Wnt pathway reference [109].

MANGANESE

The living body gets exposed to manganese through dietary, industrial, or occupational exposure as well as through inhalation in the form of manganese aerosols, along with intake in drinking water [110]. The use of manganese compounds has also been increased in magnetic resonance imaging. As anti-knock agent, manganese compounds are used in dry-cell batteries, metallurgy, dyes, ceramics, medicines, and food supplements. Over exposure of manganese has been linked to cancer and malformations [111].

Teratogenic Nature of Manganese

Studies have reported that a large amount of manganese causes infertility and abnormalities in embryo and fetus. Recently manganese compounds are used in fungicide such as MANEB and MnDPDP that are reported to be highly embryo toxic [111]. Manganese is a known developmental toxicant and an ecological association between manganese in drinking water and prevalence of congenital cardiac malformation is reported in the human population. A study in North

Carolina on manganese levels in drinking water and their effect on general population has indicated that manganese levels in well water is associated with the prevalence of birth defects such as conotruncal heart defects in the human population [112]. *In-utero* exposure of manganese to the developing fetus results in congenital malformation, abnormal neurodevelopment, reduced fetal growth, preterm birth, low birth weight, stillbirth, and spontaneous abortion [113]. Elevated placental levels of manganese result in neural tube defects mainly Spina bifida [114]. Manganese exposure is also associated with severe outcomes of pregnancy such as infant mortality, reduced birth weight, and intrauterine growth restriction.

Animal studies on rats, mice, chicks, and guinea-pigs have confirmed that manganese deficiency during the developmental period is linked with skeletal abnormalities, reduced litter size, ataxia, risk of premature birth, and stillbirth. However, studies on pregnant rodents and chick embryo treated with manganese chloride ($MnCl_2$), manganese dioxide (MnO_2) and manganese sulphate ($MnSo_4$) have indicated that increased maternal concentration of manganese not only affect fertility but also cause toxicity in mother and teratogenicity in developing fetus because it can cross the placental barrier [111]. Administration of $MnCl_2$ more than 700 µg/egg into air sacs of chick eggs on day 2 of incubation causes toxicity and teratogenic effects [115] and the dose of 9.85 ng/ml to 39.4 µg/ml is able to interfere with fetal developmental process in mice [116]. Oral supplementation of manganese 500-1000 mg/kg for up to 7.5 months in rats showed infertility and reduced growth and survival rate of neonates. Overexposure of manganese is also associated with spermatogenesis, dysfunction of sexual glands, epithelial alteration, and corpus luteum persistence [117]. Intraperitoneal injection of 12.5-50 mg/kg of MnSO4 induces exencephaly, resorption, and fetal death [118].

Mechanism of Teratogenic Activity

Increased concentration of manganese is associated with disturbances in DNA replication, chromosomal aberrations, and DNA damage. Manganese doses 5 to 6 times higher than the daily requirement in parenteral nutrition cause neurotoxicity. Risk associated with manganese toxicity and developing brain is further linked to DNA damage, increased manganese interferes with DNA integrity and replication [111]. Another suggested mechanism includes the interaction of manganese with detoxifying enzyme or redox system that protects the cell from free radical attack. Excess manganese is oxidized from Mn^{+2} to Mn^{+3}, Mn^{+3} is highly toxic in nature that facilitates the oxidation of many important biomolecules such as dopamine that further leads to the generation of several neurotoxic compounds [119].

COBALT

Cobalt is an essential trace element for many important biological functions, plays a critical role in vitamin B12 production and other cobalamine. It is usually found in the environment in different chemical forms and in association with other metals such as manganese, copper, and nickel. A small amount is beneficial for living body but higher exposure results in cobalt toxicity [120]. It is used in production of alloys, magnets, as drying agent, and as catalyst in rubber factories. Atmospheric levels of cobalt can also rise from combustion of fossil fuels and sewage sludge contamination [121].

Teratogenic Nature of Cobalt

Cobalt deficiency is associated with abnormalities in mental and physical growth of infants, however, excessive levels are found to be involved in producing harmful effects [122]. Cobalt can cross the placental barrier and accumulate in fetal tissue and amniotic fluid. Studies have reported teratogenic effects of cobalt in dose dependent manner [123]. Women using metal-on-metal hip arthroplasty having cobalt chrome alloys at child bearing age have serious concerns on developing fetus because cobalt ions are generated by metal-on metal bearing which have severe negative consequences on early embryonic development due to the toxic nature [124].

Experimental work on rats, mice, rabbits and on cell lines has reported that elevated cobalt levels in maternal blood are associated with reduce birth weight and fetal death [125]. Developmental toxicities induced by cobalt include skeletal malformations, lower body weight, and embryo lethality. Increased frequency of other anomalies of vital body parts is also reported with cobalt developmental toxicity including defects in eyes, spine, skull, kidney, and sternum in mice. It also affects fertility in time- and dose-dependent manner, decreases testicular weight, sperm motility and epididymal sperm concentration is also reported [123]. Exposure of cobalt in nematodes has been found to be involved in the reduction of growth in progeny and induces reproductive defects that are transferrable from parents to offspring [125]. Developmental toxicity is also reported in zebrafish due to cobalt exposure by inducing apoptosis and oxidative stress in developing fetus [126].

Mechanism of Teratogenic Activity

Cobalt exposure inhibits DNA repair, produces nitrogen and reactive oxygen species, and alters gene expression. It affects Fas, caspase and other signaling pathways that ultimately induce apoptosis [127]. Cobalt induces abnormalities in various biological mechanisms that are responsible for differentiation and development that ultimately interfere with developmental process. Cobalt induces developmental delay in the neuromuscular and locomotor system by interfering with presynaptic nerve ending of motor neurons [128]. It is involved in the disruption of synaptic functions and impairs the connections with motor neuron and muscles cells. Cobalt is also involved in breaking DNA strands by producing reactive oxygen species [129].

COPPER

Copper, being the essential micronutrient is required for normal developmental process. It is an important cofactor for cuproenzymes that are required for various biochemical processes such as extracellular matrix (ECM) protein crosslinking, energy production, immune function, oxidant-defense, synthesis of neuromodulators, blood cell maturation, myocardial contractility, and maintenance of iron (Fe) homeostasis. The deficiency of copper is strongly recommended to induce abnormal prenatal development in humans [130]. However, the toxicity of copper is also associated with developmental aberrations. The toxicity of copper in maternal blood usually occurs through food and drinking water or by high consumption of copper supplements [131, 132].

The direct relation between copper accumulation and developmental toxicity has been seen in human case studies. The women with Wilson's disease showed infertility due to abnormal menstruation [133]. Wilson's disease exhibits genetic mutation in copper transporter (ATP_7B) which results in incorporation of copper in ceruloplasmin and biliary secretion leading to pathological accumulation of copper in various tissues and organ including the placenta [133]. In untreated women with Wilson's disease, there is an increased chance of early abortion and if somehow the embryo survives to grow then the infant shows hepatic malfunctioning such as hepatomegaly and increased liver enzymes. These conditions are accompanied by reduced levels of ceruloplasmin and increased excretion of copper [134]. The high concentration of copper in umbilical cord and maternal blood is also found to be inversely related to birth weight in humans. The disruption of placental barrier possibly due to oxidative stress is considered as one of the reasons for the increased transportation of copper from maternal blood to fetal circulation [135].

Retarded fetal growth due to high consumption of copper has also been observed in different animal models. Increased administration of copper dose dependently resulted in growth retardation, and abnormal bone and skeletal development in rats [135, 136]. Teratogenic effects of copper have also been reported in pregnant hamsters. These copper-exposed hamsters delivered the newborn exhibiting ectopia cordis (externalization of the heart) [137]. In animal model of sheep, administration of copper through diet (25 or 50 mg/d) reduced the number of lambs born due to increased fetal abortions [138]. Copper-induced teratogenic activity is also seen in fishes. Inhabitation of fish in copper contaminated water can induce physical and sensory abnormalities such as impairment in the gills and gut, diminished ability to escape predators, and find food or locate a mate [139]. An inverse relation of increased copper concentration and delayed larval growth has also been observed in fish species of fathead minnow [140].

Mechanism of Teratogenic Activity

The teratogenicity due to copper accumulation is attributed to the elevated oxidative stress. Short-term exposure of copper to zebrafish larvae has been shown to increase the production of reactive species and depletion of endogenous antioxidant reserves. As a result, antioxidant system fails to rescue the embryo from oxidative stress-induced apoptosis. Embryonic cell apoptosis can lead to developmental malformations which may result in embryonic death [141]. Exposure to copper results in reduced activity of superoxide dismutase, catalase, and glutathione peroxidase. Inhibition of embryonic Na+/K+ATPase is also reported due to excessive exposure to copper [142, 143]. This may lead to impaired ion regulation and developmental abnormalities in the growing embryo [141, 144].

CONCLUDING REMARKS

Heavy metals exist naturally; however, the anthropogenic activities of humans also impart environmental contamination. Besides producing systemic malfunctioning, exposure to metals induces substantial damages to the growing fetus as well as post-natal abnormalities. In addition, paternal and maternal exposures to metals are also associated with infertility and/or increased cases of miscarriages (Fig. 5). Daily exposure to metals by different sources including occupation, food, drinking water, pollution, and smoking, has proportionally increased the chances of developmental toxicity in human population. The metals which are discussed in this chapter are shown to have a teratogenic property in various clinical and pre-clinical studies. The basic mechanism of metal-induced toxicity involves oxidative stress, inflammation, and apoptosis, however, developmental toxicity of all metals is not fully elucidated. It is, therefore, necessary to conduct experimental studies to understand the mechanism of action of teratogenicity of each metal in order to develop an effective treatment that would help to reduce toxicological effects of metals during development. Moreover, the cases of infertility, early embryonic death, pregnancy complications, pre- and post-natal abnormalities have shown to be directly associated with different metal concentrations, therefore, the especial concern should be given for the detection of metal levels while dealing with such cases. Most importantly, the environmental contamination with different metals should be minimized so that the living organism and thus the entire food chain become protected from the metal exposure. Especial preventive measures should be taken during developmental process to avoid metal toxicity-induced embryopathogenesis. It is also necessary to define guidelines and rules by the law-makers to identify the areas with higher levels of metal contamination in the environment. Otherwise, failure to do so will impose severe complications in the future generations.

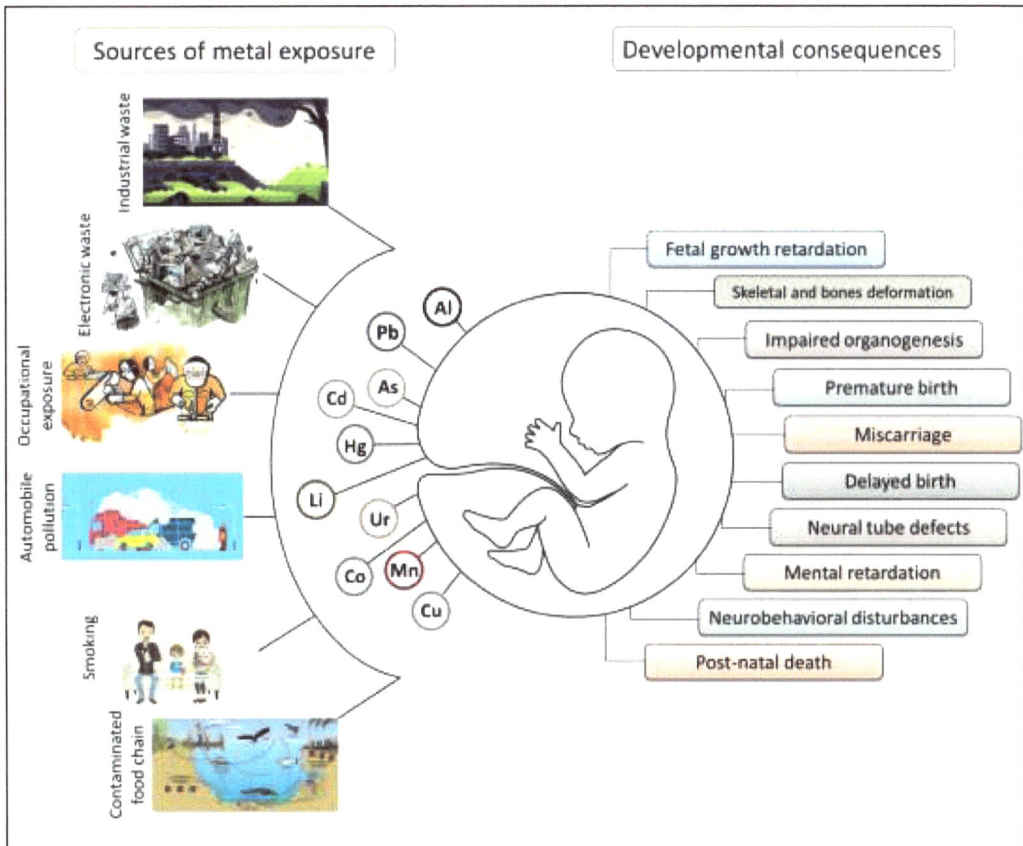

Fig. (5). Environmental dissemination of different metals because of anthropogenic activities of humans results in several developmental abnormalities.

CONSENT FOR PUBLICATION

Not applicable.

CONFLICT OF INTEREST

The author declares no conflict of interest, financial or otherwise.

ACKNOWLEDGEMENTS

Declared none

REFERENCES

[1]   Abd-Elhady RM, Elsheikh AM, Khalifa AE. Anti-amnestic properties of Ginkgo biloba extract on impaired memory function induced by Aluminium in rats. Int J Dev Neurosci 2013; 31(7): 598-607.
[http://dx.doi.org/10.1016/j.ijdevneu.2013.07.006] [PMID: 23933390]

[2]   Abu-Taweel GM, Ajarem JS, Ahmad M. Neurobehavioral toxic effects of perinatal oral exposure to Aluminium on the developmental motor reflexes, learning, memory and brain neurotransmitters of mice offspring. Pharmacol Biochem Behav 2012; 101(1): 49-56.
[http://dx.doi.org/10.1016/j.pbb.2011.11.003] [PMID: 22115621]

[3]   Rollin H, Channa K, Olutola B, Nogueira C, Odland J. Prenatal exposure to aluminium in South Africa: An emerging concern. Environ Epidemiol 2019; 3: 338.
[http://dx.doi.org/10.1097/01.EE9.0000609720.08410.8c]

[4]   Sharma P, Mishra KP. Aluminium-induced maternal and developmental toxicity and oxidative stress in rat brain: response to combined administration of Tiron and glutathione. Reprod Toxicol 2006; 21(3): 313-21.
[http://dx.doi.org/10.1016/j.reprotox.2005.06.004] [PMID: 16040227]

[5]   Reinke CM, Breitkreutz J, Leuenberger H. Aluminium in over-the-counter drugs: risks outweigh benefits? Drug Saf 2003; 26(14): 1011-25.
[http://dx.doi.org/10.2165/00002018-200326140-00003] [PMID: 14583063]

[6]   Bellés M, Albina ML, Sánchez DJ, Domingo JL. Lack of protective effects of dietary silicon on aluminium-induced maternal and developmental toxicity in mice. Pharmacol Toxicol 1999; 85(1): 1-6.
[http://dx.doi.org/10.1111/j.1600-0773.1999.tb01055.x] [PMID: 10426156]

[7]   Khalaf AA, Morgan AM, Mekawy MM, Ali MF. Developmental toxicity evaluation of oral Aluminium in rats. J Egypt Soc Toxicol 2007; 37: 11-26.

[8]   Colomina MT, Roig JL, Sánchez DJ, Domingo JL. Influence of age on Aluminium-induced neurobehavioral effects and morphological changes in rat brain. Neurotoxicology 2002; 23(6): 775-81.
[http://dx.doi.org/10.1016/S0161-813X(02)00008-6] [PMID: 12520767]

[9]   Yumoto S, Nagai H, Matsuzaki H, *et al.* Aluminium incorporation into the brain of rat fetuses and sucklings. Brain Res Bull 2001; 55(2): 229-34.
[http://dx.doi.org/10.1016/S0361-9230(01)00509-3] [PMID: 11470320]

[10] Bellés M, Albina ML, Sanchez DJ, Corbella J, Domingo JL. Effects of oral Aluminium on essential trace elements metabolism during pregnancy. Biol Trace Elem Res 2001; 79(1): 67-81.
[http://dx.doi.org/10.1385/BTER:79:1:67] [PMID: 11318238]

[11] Schwalfenberg GK, Genuis SJ. Vitamin D, essential minerals, and toxic elements: exploring interactions between nutrients and toxicants in clinical medicine. ScientificWorldJournal 2015; 2015: 318595.
[http://dx.doi.org/10.1155/2015/318595] [PMID: 26347061]

[12] González-Suárez I, Alvarez-Hernández D, Carrillo-López N, Naves-Díaz M, Luis Fernández-Martín J, Cannata-Andía JB. Aluminium posttranscriptional regulation of parathyroid hormone synthesis: a role for the calcium-sensing receptor. Kidney Int 2005; 68(6): 2484-96.
[http://dx.doi.org/10.1111/j.1523-1755.2005.00724.x] [PMID: 16316325]

[13]   Hill DS, Wlodarczyk BJ, Finnell RH. Reproductive consequences of oral arsenate exposure during pregnancy in a mouse model. Birth Defects Res B Dev Reprod Toxicol 2008; 83(1): 40-7.
[http://dx.doi.org/10.1002/bdrb.20142] [PMID: 18186108]

[14]   Mazumdar M. Does arsenic increase the risk of neural tube defects among a highly exposed population? A new case-control study in Bangladesh. Birth Defects Res 2017; 109(2): 92-8.
[http://dx.doi.org/10.1002/bdra.23577] [PMID: 27801974]

[15]   Devesa V, Adair BM, Liu J, *et al.* Arsenicals in maternal and fetal mouse tissues after gestational exposure to arsenite. Toxicology 2006; 224(1-2): 147-55.
[http://dx.doi.org/10.1016/j.tox.2006.04.041] [PMID: 16753250]

[16]   Carter DE, Aposhian HV, Gandolfi AJ. The metabolism of inorganic arsenic oxides, gallium arsenide, and arsine: a toxicochemical review. Toxicol Appl Pharmacol 2003; 193(3): 309-34.
[http://dx.doi.org/10.1016/j.taap.2003.07.009] [PMID: 14678742]

[17]   Milton AH, Smith W, Rahman B, *et al.* Chronic arsenic exposure and adverse pregnancy outcomes in bangladesh. Epidemiology 2005; 16(1): 82-6.
[http://dx.doi.org/10.1097/01.ede.0000147105.94041.e6] [PMID: 15613949]

[18]   Parvez F, Wasserman GA, Factor-Litvak P, *et al.* Arsenic exposure and motor function among children in Bangladesh. Environ Health Perspect 2011; 119(11): 1665-70.
[http://dx.doi.org/10.1289/ehp.1103548] [PMID: 21742576]

[19]   Hamadani JD, Grantham-McGregor SM, Tofail F, *et al.* Pre- and postnatal arsenic exposure and child development at 18 months of age: a cohort study in rural Bangladesh. Int J Epidemiol 2010; 39(5): 1206-16.
[http://dx.doi.org/10.1093/ije/dyp369] [PMID: 20085967]

[20]   Rodrigues EG, Bellinger DC, Valeri L, *et al.* Neurodevelopmental outcomes among 2- to 3-year-old children in Bangladesh with elevated blood lead and exposure to arsenic and manganese in drinking water. Environ Health 2016; 15(1): 44.
[http://dx.doi.org/10.1186/s12940-016-0127-y] [PMID: 26968381]

[21]   Willhite CC. Arsenic-induced axial skeletal (dysraphic) disorders. Exp Mol Pathol 1981; 34(2): 145-58.
[http://dx.doi.org/10.1016/0014-4800(81)90071-X] [PMID: 6894125]

[22]   Yang P, Li X, Xu C, *et al.* Maternal hyperglycemia activates an ASK1-FoxO3a-caspase 8 pathway that leads to embryonic neural tube defects. Sci Signal 2013; 6(290): ra74.
[http://dx.doi.org/10.1126/scisignal.2004020] [PMID: 23982205]

[23]   Gandhi DN, Kumar R. Arsenic toxicity and neurobehaviors: a review. Inn Pharma Pharmacotherapy 2013; 1(1): 1-5.

[24]   Gefrides LA, Bennett GD, Finnell RH. Effects of folate supplementation on the risk of spontaneous and induced neural tube defects in Splotch mice. Teratology 2002; 65(2): 63-9.
[http://dx.doi.org/10.1002/tera.10019] [PMID: 11857507]

[25]   Mitchell LE. Epidemiology of neural tube defects. Am J Med Genet C Semin Med Genet 2005; 135C(1): 88-94.
[http://dx.doi.org/10.1002/ajmg.c.30057] [PMID: 15800877]

[26]   Kovacic P, Somanathan R. Mechanism of teratogenesis: electron transfer, reactive oxygen species, and antioxidants. Birth Defects Res C Embryo Today 2006; 78(4): 308-25.
[http://dx.doi.org/10.1002/bdrc.20081] [PMID: 17315244]

[27]   Várnagy L, Budai P, Molnár E, Takács I, Kárpáti A. Interaction of Dithane M-45 (mancozeb) and lead acetate during a teratogenicity test in rats. Acta Vet Hung 2000; 48(1): 113-24.
[http://dx.doi.org/10.1556/avet.48.2000.1.13] [PMID: 11402670]

[28]   Jabeen R, Tahir M, Waqas S. Teratogenic effects of lead acetate on kidney. J Ayub Med Coll

Abbottabad 2010; 22(1): 76-9.
[PMID: 21409910]

[29]   Lavolpe M, Greco LL, Kesselman D, Rodríguez E. Differential toxicity of copper, zinc, and lead during the embryonic development of Chasmagnathus granulatus (Brachyura, Varunidae). Environ Toxicol Chem 2004; 23(4): 960-7.
[http://dx.doi.org/10.1897/02-645] [PMID: 15095892]

[30]   Weizsaecker K. Lead toxicity during pregnancy. Prim Care Update Ob Gyns 2003; 10(6): 304-9.
[http://dx.doi.org/10.1016/S1068-607X(03)00074-X]

[31]   Bound JP, Harvey PW, Francis BJ, Awwad F, Gatrell AC. Involvement of deprivation and environmental lead in neural tube defects: a matched case-control study. Arch Dis Child 1997; 76(2): 107-12.
[http://dx.doi.org/10.1136/adc.76.2.107] [PMID: 9068297]

[32]   Bokara KK, Brown E, McCormick R, Yallapragada PR, Rajanna S, Bettaiya R. Lead-induced increase in antioxidant enzymes and lipid peroxidation products in developing rat brain. Biometals 2008; 21(1): 9-16.
[http://dx.doi.org/10.1007/s10534-007-9088-5] [PMID: 18214713]

[33]   Irgens A, Krüger K, Skorve AH, Irgens LM. Reproductive outcome in offspring of parents occupationally exposed to lead in Norway. Am J Ind Med 1998; 34(5): 431-7.
[http://dx.doi.org/10.1002/(SICI)1097-0274(199811)34:5<431::AID-AJIM3>3.0.CO;2-T] [PMID: 9787846]

[34]   Brender JD, Suarez L, Felkner M, *et al.* Maternal exposure to arsenic, cadmium, lead, and mercury and neural tube defects in offspring. Environ Res 2006; 101(1): 132-9.
[http://dx.doi.org/10.1016/j.envres.2005.08.003] [PMID: 16171797]

[35]   Cengiz B, Söylemez F, Oztürk E, Çavdar AO. Serum zinc, selenium, copper, and lead levels in women with second-trimester induced abortion resulting from neural tube defects: a preliminary study. Biol Trace Elem Res 2004; 97(3): 225-35.
[http://dx.doi.org/10.1385/BTER:97:3:225] [PMID: 14997023]

[36]   Hu H, Téllez-Rojo MM, Bellinger D, *et al.* Fetal lead exposure at each stage of pregnancy as a predictor of infant mental development. Environ Health Perspect 2006; 114(11): 1730-5.
[http://dx.doi.org/10.1289/ehp.9067] [PMID: 17107860]

[37]   Bellinger DC. Teratogen update: lead and pregnancy. Birth Defects Res A Clin Mol Teratol 2005; 73(6): 409-20.
[http://dx.doi.org/10.1002/bdra.20127] [PMID: 15880700]

[38]   Hertz-Picciotto I, Schramm M, Watt-Morse M, Chantala K, Anderson J, Osterloh J. Patterns and determinants of blood lead during pregnancy. Am J Epidemiol 2000; 152(9): 829-37.
[http://dx.doi.org/10.1093/aje/152.9.829] [PMID: 11085394]

[39]   Falcón M, Viñas P, Luna A. Placental lead and outcome of pregnancy. Toxicology 2003; 185(1-2): 59-66.
[http://dx.doi.org/10.1016/S0300-483X(02)00589-9] [PMID: 12505445]

[40]   Sallmén M, Lindbohm ML, Anttila A, Taskinen H, Hemminki K. Paternal occupational lead exposure and congenital malformations. J Epidemiol Community Health 1992; 46(5): 519-22.
[http://dx.doi.org/10.1136/jech.46.5.519] [PMID: 1479323]

[41]   Osman AG, Wuertz S, Mekkawy IA, Exner HJ, Kirschbaum F. Lead induced malformations in embryos of the African catfish Clarias gariepinus (Burchell, 1822). Environ Toxicol 2007; 22(4): 375-89.
[http://dx.doi.org/10.1002/tox.20272] [PMID: 17607729]

[42]   Ronis MJ, Gandy J, Badger T. Endocrine mechanisms underlying reproductive toxicity in the developing rat chronically exposed to dietary lead. J Toxicol Environ Health A 1998; 54(2): 77-99.

[http://dx.doi.org/10.1080/009841098158935] [PMID: 9652546]

[43]   Junaid M, Chowdhuri DK, Narayan R, Shanker R, Saxena DK. Lead-induced changes in ovarian follicular development and maturation in mice. J Toxicol Environ Health 1997; 50(1): 31-40.
[http://dx.doi.org/10.1080/009841097160582] [PMID: 9015130]

[44]   Naha N, Bhar RB, Mukherjee A, Chowdhury AR. Structural alteration of spermatozoa in the persons employed in lead acid battery factory. Indian J Physiol Pharmacol 2005; 49(2): 153-62.
[PMID: 16170983]

[45]   Flora JSW, Agrawal S. Arsenic, cadmium, and lead. In: Gupta RC, Ed. Reproductive and developmental toxicology. 2nd ed. San Diego, CA: Elsevier, Academic Press 2017; pp. 537-66.
[http://dx.doi.org/10.1016/B978-0-12-804239-7.00031-7]

[46]   Sokol RZ, Wang S, Wan YJ, Stanczyk FZ, Gentzschein E, Chapin RE. Long-term, low-dose lead exposure alters the gonadotropin-releasing hormone system in the male rat. Environ Health Perspect 2002; 110(9): 871-4.
[http://dx.doi.org/10.1289/ehp.02110871] [PMID: 12204820]

[47]   Zhang H, Liu Y, Zhang R, Liu R, Chen Y. Binding mode investigations on the interaction of lead(II) acetate with human chorionic gonadotropin. J Phys Chem B 2014; 118(32): 9644-50.
[http://dx.doi.org/10.1021/jp505565s] [PMID: 25096834]

[48]   Liu J, Shi JZ, Yu LM, Goyer RA, Waalkes MP. Mercury in traditional medicines: is cinnabar toxicologically similar to common mercurials? Exp Biol Med (Maywood) 2008; 233(7): 810-7.
[http://dx.doi.org/10.3181/0712-MR-336] [PMID: 18445765]

[49]   Clarkson TW, Magos L, Myers GJ. The toxicology of mercury--current exposures and clinical manifestations. N Engl J Med 2003; 349(18): 1731-7.
[http://dx.doi.org/10.1056/NEJMra022471] [PMID: 14585942]

[50]   Morgan DL, Chanda SM, Price HC, *et al.* Disposition of inhaled mercury vapor in pregnant rats: maternal toxicity and effects on developmental outcome. Toxicol Sci 2002; 66(2): 261-73.
[http://dx.doi.org/10.1093/toxsci/66.2.261] [PMID: 11896293]

[51]   Koos BJ, Longo LD. Mercury toxicity in the pregnant woman, fetus, and newborn infant. A review. Am J Obstet Gynecol 1976; 126(3): 390-409.
[http://dx.doi.org/10.1016/0002-9378(76)90557-3] [PMID: 786026]

[52]   Myers GJ, Davidson PW, Cox C, *et al.* Prenatal methylmercury exposure from ocean fish consumption in the Seychelles child development study. Lancet 2003; 361(9370): 1686-92.
[http://dx.doi.org/10.1016/S0140-6736(03)13371-5] [PMID: 12767734]

[53]   Grandjean P, Weihe P. Neurobehavioral effects of intrauterine mercury exposure: potential sources of bias. Environ Res 1993; 61(1): 176-83.
[http://dx.doi.org/10.1006/enrs.1993.1062] [PMID: 8472672]

[54]   Schuurs AH. Reproductive toxicity of occupational mercury. A review of the literature. J Dent 1999; 27(4): 249-56.
[http://dx.doi.org/10.1016/S0300-5712(97)00039-0] [PMID: 10193101]

[55]   Baranski B, Szymczyk I. Effects of mercury vapor on reproductive functions of female white rats. Med Pr 1973; 24: 249-61.

[56]   Dong W, Liu J, Wei L, Jingfeng Y, Chernick M, Hinton DE. Developmental toxicity from exposure to various forms of mercury compounds in medaka fish (Oryzias latipes) embryos. PeerJ 2016; 4: e2282.
[http://dx.doi.org/10.7717/peerj.2282] [PMID: 27635309]

[57]   Domingo JL. Metal-induced developmental toxicity in mammals: a review. J Toxicol Environ Health 1994; 42(2): 123-41.
[http://dx.doi.org/10.1080/15287399409531868] [PMID: 8207750]

[58]   Ishitobi H, Stern S, Thurston SW, *et al.* Organic and inorganic mercury in neonatal rat brain after

prenatal exposure to methylmercury and mercury vapor. Environ Health Perspect 2010; 118(2): 242-8.
[http://dx.doi.org/10.1289/ehp.0900956] [PMID: 20123608]

[59]　Magos L, Halbach S, Clarkson TW. Role of catalase in the oxidation of mercury vapor. Biochem Pharmacol 1978; 27(9): 1373-7.
[http://dx.doi.org/10.1016/0006-2952(78)90122-3] [PMID: 567993]

[60]　Moynihan M, Peterson KE, Cantoral A, *et al.* Dietary predictors of urinary cadmium among pregnant women and children. Sci Total Environ 2017; 575: 1255-62.
[http://dx.doi.org/10.1016/j.scitotenv.2016.09.204] [PMID: 27707662]

[61]　Piasek M, Blanuša M, Kostial K, Laskey JW. Placental cadmium and progesterone concentrations in cigarette smokers. Reprod Toxicol 2001; 15(6): 673-81.
[http://dx.doi.org/10.1016/S0890-6238(01)00174-5] [PMID: 11738520]

[62]　Piasek M, Laskey JW, Kostial K, Blanuša M. Assessment of steroid disruption using cultures of whole ovary and/or placenta in rat and in human placental tissue. Int Arch Occup Environ Health 2002; 75(1) (Suppl.): S36-44.
[http://dx.doi.org/10.1007/s00420-002-0351-3] [PMID: 12397409]

[63]　Sukhodolska N. Comparative analysis of toxic (cadmium, lead) and trace (zinc, copper) elements content in women's blood during I trimester of uncomplicated and complicated course of gestation. Proc Shevchenko Sci Soc Med Sci 2017; 49(1): 64-73.

[64]　Taylor CM, Emond AM, Lingam R, Golding J. Prenatal lead, cadmium and mercury exposure and associations with motor skills at age 7 years in a UK observational birth cohort. Environ Int 2018; 117: 40-7.
[http://dx.doi.org/10.1016/j.envint.2018.04.032] [PMID: 29723752]

[65]　Halder S, Kar R, Galav V, *et al.* Cadmium exposure during lactation causes learning and memory-impairment in F1 generation mice: amelioration by quercetin. Drug Chem Toxicol 2016; 39(3): 272-8.
[http://dx.doi.org/10.3109/01480545.2015.1092042] [PMID: 26446883]

[66]　Dharmadasa P, Kim N, Thunders M. Maternal cadmium exposure and impact on foetal gene expression through methylation changes. Food Chem Toxicol 2017; 109(Pt 1): 714-20.
[http://dx.doi.org/10.1016/j.fct.2017.09.002] [PMID: 28887092]

[67]　Akinloye O, Arowojolu AO, Shittu OB, Anetor JI. Cadmium toxicity: a possible cause of male infertility in Nigeria. Reprod Biol 2006; 6(1): 17-30.
[PMID: 16604149]

[68]　Goyer RA, Liu J, Waalkes MP. Cadmium and cancer of prostate and testis. Biometals 2004; 17(5): 555-8.
[http://dx.doi.org/10.1023/B:BIOM.0000045738.59708.20] [PMID: 15688863]

[69]　Sunderman FW Jr, Plowman MC, Hopfer SM. Teratogenicity of cadmium chloride in the South African frog, Xenopus laevis. IARC Sci Publ 1992; 1(118): 249-56.
[PMID: 1303948]

[70]　Thompson J, Bannigan J. Cadmium: toxic effects on the reproductive system and the embryo. Reprod Toxicol 2008; 25(3): 304-15.
[http://dx.doi.org/10.1016/j.reprotox.2008.02.001] [PMID: 18367374]

[71]　Menoud PA, Schowing J. A preliminary study of the mechanisms of cadmium teratogenicity in chick embryo after direct action. J Toxicol Clin Exp 1987; 7(2): 77-84.
[PMID: 3656209]

[72]　El-Demerdash FM, Yousef MI, Kedwany FS, Baghdadi HH. Cadmium-induced changes in lipid peroxidation, blood hematology, biochemical parameters and semen quality of male rats: protective role of vitamin E and β-carotene. Food Chem Toxicol 2004; 42(10): 1563-71.
[http://dx.doi.org/10.1016/j.fct.2004.05.001] [PMID: 15304303]

[73]　Waalkes MP, Rehm S, Devor DE. The effects of continuous testosterone exposure on spontaneous and

cadmium-induced tumors in the male Fischer (F344/NCr) rat: loss of testicular response. Toxicol Appl Pharmacol 1997; 142(1): 40-6.
[http://dx.doi.org/10.1006/taap.1996.8005] [PMID: 9007032]

[74]   Bitto A, Pizzino G, Irrera N, Galfo F, Squadrito F. Epigenetic modifications due to heavy metals exposure in children living in polluted areas. Curr Genomics 2014; 15(6): 464-8.
[http://dx.doi.org/10.2174/1389202915066150106153336] [PMID: 25646074]

[75]   Vilahur N, Vahter M, Broberg K. The epigenetic effects of prenatal cadmium exposure. Curr Environ Health Rep 2015; 2(2): 195-203.
[http://dx.doi.org/10.1007/s40572-015-0049-9] [PMID: 25960943]

[76]   Geng HX, Wang L. Cadmium: Toxic effects on placental and embryonic development. Environ Toxicol Pharmacol 2019; 67: 102-7.
[http://dx.doi.org/10.1016/j.etap.2019.02.006] [PMID: 30797179]

[77]   Kippler M, Engström K, Mlakar SJ, *et al.* Sex-specific effects of early life cadmium exposure on DNA methylation and implications for birth weight. Epigenetics 2013; 8(5): 494-503.
[http://dx.doi.org/10.4161/epi.24401] [PMID: 23644563]

[78]   Everson TM, Armstrong DA, Jackson BP, Green BB, Karagas MR, Marsit CJ. Maternal cadmium, placental PCDHAC1, and fetal development. Reprod Toxicol 2016; 65: 263-71.
[http://dx.doi.org/10.1016/j.reprotox.2016.08.011] [PMID: 27544570]

[79]   Appleton AA, Jackson BP, Karagas M, Marsit CJ. Prenatal exposure to neurotoxic metals is associated with increased placental glucocorticoid receptor DNA methylation. Epigenetics 2017; 12(8): 607-15.
[http://dx.doi.org/10.1080/15592294.2017.1320637] [PMID: 28548590]

[80]   Vaswani K, Dekker Nitert M, Chan HW, *et al.* Mid-to-late gestational changes in inflammatory gene expression in the rat placenta. Reprod Sci 2018; 25(2): 222-9.
[http://dx.doi.org/10.1177/1933719117741375] [PMID: 29153059]

[81]   Cotechini T, Komisarenko M, Sperou A, Macdonald-Goodfellow S, Adams MA, Graham CH. Inflammation in rat pregnancy inhibits spiral artery remodeling leading to fetal growth restriction and features of preeclampsia. J Exp Med 2014; 211(1): 165-79.
[http://dx.doi.org/10.1084/jem.20130295] [PMID: 24395887]

[82]   Liang T, Jinglong X, Shusheng D, Aiyou W. Maternal obesity stimulates lipotoxicity and up-regulates inflammatory signaling pathways in the full-term swine placenta. Anim Sci J 2018; 89(9): 1310-22.
[http://dx.doi.org/10.1111/asj.13064] [PMID: 29947166]

[83]   Zenerino C, Nuzzo AM, Giuffrida D, *et al.* The HMGB1/RAGE pro-inflammatory axis in the human placenta: Modulating effect of low molecular weight heparin. Molecules 2017; 22(11): 1997.
[http://dx.doi.org/10.3390/molecules22111997] [PMID: 29149067]

[84]   Hu J, Wang H, Hu YF, *et al.* Cadmium induces inflammatory cytokines through activating Akt signaling in mouse placenta and human trophoblast cells. Placenta 2018; 65: 7-14.
[http://dx.doi.org/10.1016/j.placenta.2018.03.008] [PMID: 29908644]

[85]   Guo J, Wu C, Qi X, *et al.* Adverse associations between maternal and neonatal cadmium exposure and birth outcomes. Sci Total Environ 2017; 575: 581-7.
[http://dx.doi.org/10.1016/j.scitotenv.2016.09.016] [PMID: 27614860]

[86]   Mikolić A, Piasek M, Sulimanec Grgec A, Varnai VM, Stasenko S, Kralik Oguić S. Oral cadmium exposure during rat pregnancy: assessment of transplacental micronutrient transport and steroidogenesis at term. J Appl Toxicol 2015; 35(5): 508-19.
[http://dx.doi.org/10.1002/jat.3055] [PMID: 25256609]

[87]   Chowanadisai W, Graham DM, Keen CL, Rucker RB, Messerli MA. Neurulation and neurite extension require the zinc transporter ZIP12 (slc39a12). Proc Natl Acad Sci USA 2013; 110(24): 9903-8.
[http://dx.doi.org/10.1073/pnas.1222142110] [PMID: 23716681]

[88]   Wang H, Wang Y, Bo QL, *et al.* Maternal cadmium exposure reduces placental zinc transport and induces fetal growth restriction in mice. Reprod Toxicol 2016; 63: 174-82.
[http://dx.doi.org/10.1016/j.reprotox.2016.06.010] [PMID: 27319394]

[89]   Hindin R, Brugge D, Panikkar B. Teratogenicity of depleted uranium aerosols: a review from an epidemiological perspective. Environ Health 2005; 4(1): 17.
[http://dx.doi.org/10.1186/1476-069X-4-17] [PMID: 16124873]

[90]   Bleise A, Danesi PR, Burkart W. Properties, use and health effects of depleted uranium (DU): a general overview. J Environ Radioact 2003; 64(2-3): 93-112.
[http://dx.doi.org/10.1016/S0265-931X(02)00041-3] [PMID: 12500797]

[91]   Duraković A. Undiagnosed illnesses and radioactive warfare. Croat Med J 2003; 44(5): 520-32.
[PMID: 14515407]

[92]   Hassan AA, Ghieth BM, Khalil AF, Nigm AA. Uranium favourability and its mode of migration in some rocks in the south eastern desert, Egypt. Egypt J Geo 2003; 47: 2.

[93]   Al-Shammosy MM.

[94]   Araneta MRG, Destiche DA, Schlangen KM, Merz RD, Forrester MB, Gray GC. Birth defects prevalence among infants of Persian Gulf War veterans born in Hawaii, 1989-1993. Teratology 2000; 62(4): 195-204.
[http://dx.doi.org/10.1002/1096-9926(200010)62:4<195::AID-TERA5>3.0.CO;2-5]   [PMID: 10992261]

[95]   Araneta MR, Schlangen KM, Edmonds LD, *et al.* Prevalence of birth defects among infants of Gulf War veterans in Arkansas, Arizona, California, Georgia, Hawaii, and Iowa, 1989-1993. Birth Defects Res A Clin Mol Teratol 2003; 67(4): 246-60.
[http://dx.doi.org/10.1002/bdra.10033] [PMID: 12854660]

[96]   Domingo JL. Reproductive and developmental toxicity of natural and depleted uranium: a review. Reprod Toxicol 2001; 15(6): 603-9.
[http://dx.doi.org/10.1016/S0890-6238(01)00181-2] [PMID: 11738513]

[97]   Horton S, Tuerk A, Cook D, Cook J, Dhurjati P. Maximum recommended dosage of lithium for pregnant women based on a PBPK model for lithium absorption. Adv Bioinforma 2012; 2012: 352729.
[http://dx.doi.org/10.1155/2012/352729] [PMID: 22693500]

[98]   Kozma C. Neonatal toxicity and transient neurodevelopmental deficits following prenatal exposure to lithium: Another clinical report and a review of the literature. Am J Med Genet A 2005; 132A(4): 441-4.
[http://dx.doi.org/10.1002/ajmg.a.30501] [PMID: 15633173]

[99]   Nokhbatolfoghahai M, Parivar K. Teratogenic effect of lithium carbonate in early development of BALB/c mouse. Anat Rec (Hoboken) 2008; 291(9): 1088-96.
[http://dx.doi.org/10.1002/ar.20730] [PMID: 18727075]

[100] Larsen ER, Damkier P, Pedersen LH, *et al.* Use of psychotropic drugs during pregnancy and breast-feeding. Acta Psychiatr Scand Suppl 2015; 132(445): 1-28.
[http://dx.doi.org/10.1111/acps.12479] [PMID: 26344706]

[101] Poels EMP, Bijma HH, Galbally M, Bergink V. Lithium during pregnancy and after delivery: a review. Int J Bipolar Disord 2018; 6(1): 26.
[http://dx.doi.org/10.1186/s40345-018-0135-7] [PMID: 30506447]

[102] Patorno E, Huybrechts KF, Bateman BT, *et al.* Lithium use in pregnancy and the risk of cardiac malformations. N Engl J Med 2017; 376(23): 2245-54.
[http://dx.doi.org/10.1056/NEJMoa1612222] [PMID: 28591541]

[103] Diav-Citrin O, Shechtman S, Tahover E, *et al.* Pregnancy outcome following in utero exposure to

lithium: a prospective, comparative, observational study. Am J Psychiatry 2014; 171(7): 785-94.
[http://dx.doi.org/10.1176/appi.ajp.2014.12111402] [PMID: 24781368]

[104] Giles JJ, Bannigan JG. Teratogenic and developmental effects of lithium. Curr Pharm Des 2006; 12(12): 1531-41.
[http://dx.doi.org/10.2174/138161206776389804] [PMID: 16611133]

[105] Schlotz W, Phillips DI. Fetal origins of mental health: evidence and mechanisms. Brain Behav Immun 2009; 23(7): 905-16.
[http://dx.doi.org/10.1016/j.bbi.2009.02.001] [PMID: 19217937]

[106] Blake LD, Lucas DN, Aziz K, Castello-Cortes A, Robinson PN. Lithium toxicity and the parturient: case report and literature review. Int J Obstet Anesth 2008; 17(2): 164-9.
[http://dx.doi.org/10.1016/j.ijoa.2007.09.014] [PMID: 18308554]

[107] Corada M, Nyqvist D, Orsenigo F, *et al.* The Wnt/β-catenin pathway modulates vascular remodeling and specification by upregulating Dll4/Notch signaling. Dev Cell 2010; 18(6): 938-49.
[http://dx.doi.org/10.1016/j.devcel.2010.05.006] [PMID: 20627076]

[108] Jope RS. Lithium and GSK-3: one inhibitor, two inhibitory actions, multiple outcomes. Trends Pharmacol Sci 2003; 24(9): 441-3.
[http://dx.doi.org/10.1016/S0165-6147(03)00206-2] [PMID: 12967765]

[109] Pakula H, Xiang D, Li Z. A Tale of Two Signals: AR and WNT in Development and Tumorigenesis of Prostate and Mammary Gland. Cancers (Basel) 2017; 9(2): 14.
[http://dx.doi.org/10.3390/cancers9020014] [PMID: 28134791]

[110] Bouchard MF, Sauvé S, Barbeau B, *et al.* Intellectual impairment in school-age children exposed to manganese from drinking water. Environ Health Perspect 2011; 119(1): 138-43.
[http://dx.doi.org/10.1289/ehp.1002321] [PMID: 20855239]

[111] Gerber GB, Léonard A, Hantson P. Carcinogenicity, mutagenicity and teratogenicity of manganese compounds. Crit Rev Oncol Hematol 2002; 42(1): 25-34.
[http://dx.doi.org/10.1016/S1040-8428(01)00178-0] [PMID: 11923066]

[112] Sanders AP, Desrosiers TA, Warren JL, *et al.* Association between arsenic, cadmium, manganese, and lead levels in private wells and birth defects prevalence in North Carolina: a semi-ecologic study. BMC Public Health 2014; 14(1): 955.
[http://dx.doi.org/10.1186/1471-2458-14-955] [PMID: 25224535]

[113] Yu XD, Yan CH, Shen XM, *et al.* Prenatal exposure to multiple toxic heavy metals and neonatal neurobehavioral development in Shanghai, China. Neurotoxicol Teratol 2011; 33(4): 437-43.
[http://dx.doi.org/10.1016/j.ntt.2011.05.010] [PMID: 21664460]

[114] Liu J, Jin L, Zhang L, *et al.* Placental concentrations of manganese and the risk of fetal neural tube defects. J Trace Elem Med Biol 2013; 27(4): 322-5.
[http://dx.doi.org/10.1016/j.jtemb.2013.04.001] [PMID: 23664920]

[115] Gilani SH, Alibhai Y. Teratogenicity of metals to chick embryos. J Toxicol Environ Health 1990; 30(1): 23-31.
[http://dx.doi.org/10.1080/15287399009531407] [PMID: 2348478]

[116] Hanna LA, Peters JM, Wiley LM, Clegg MS, Keen CL. Comparative effects of essential and nonessential metals on preimplantation mouse embryo development *in vitro*. Toxicology 1997; 116(1-3): 123-31.
[http://dx.doi.org/10.1016/S0300-483X(96)03534-2] [PMID: 9020513]

[117] Corella Vargas R. [High levels of manganese in the diet of rats (Rattus norvegicus albinicus). I. Effect on reproduction]. Arch Latinoam Nutr 1984; 34(3): 457-65.
[PMID: 6544055]

[118] Webster WS, Valois AA. Reproductive toxicology of manganese in rodents, including exposure during the postnatal period. Neurotoxicology 1987; 8(3): 437-44.

[PMID: 3658242]

[119] Lima PD, Vasconcellos MC, Montenegro RC, *et al.* Genotoxic effects of Aluminium, iron and manganese in human cells and experimental systems: a review of the literature. Hum Exp Toxicol 2011; 30(10): 1435-44.
[http://dx.doi.org/10.1177/0960327110396531] [PMID: 21247993]

[120] Lukac N, Massanyi P, Zakrzewski M, Toman R, Cigankova V, Stawarz R. Cobalt-induced alterations in hamster testes in vivo. J Environ Sci Health Part A Tox Hazard Subst Environ Eng 2007; 42(3): 389-92.
[http://dx.doi.org/10.1080/10934520601144709] [PMID: 17365306]

[121] Barceloux DG, Barceloux D. Cobalt. J Toxicol Clin Toxicol 1999; 37(2): 201-6.
[http://dx.doi.org/10.1081/CLT-100102420] [PMID: 10382556]

[122] Stabler SP, Allen RH. Vitamin B12 deficiency as a worldwide problem. Annu Rev Nutr 2004; 24: 299-326.
[http://dx.doi.org/10.1146/annurev.nutr.24.012003.132440] [PMID: 15189123]

[123] Pedigo NG, George WJ, Anderson MB. Effects of acute and chronic exposure to cobalt on male reproduction in mice. Reprod Toxicol 1988; 2(1): 45-53.
[http://dx.doi.org/10.1016/S0890-6238(88)80008-X] [PMID: 2980401]

[124] Oppermann M, Borisch C, Schaefer C. Hip arthroplasty with high chromium and cobalt blood levels--Case report of a patient followed during pregnancy and lactation period. Reprod Toxicol 2015; 53: 51-3.
[http://dx.doi.org/10.1016/j.reprotox.2015.03.009] [PMID: 25828057]

[125] Wang Y, Xie W, Wang D. Transferable properties of multi-biological toxicity caused by cobalt exposure in Caenorhabditis elegans. Environ Toxicol Chem 2007; 26(11): 2405-12.
[http://dx.doi.org/10.1897/06-646R1.1] [PMID: 17941732]

[126] Cai G, Zhu J, Shen C, Cui Y, Du J, Chen X. The effects of cobalt on the development, oxidative stress, and apoptosis in zebrafish embryos. Biol Trace Elem Res 2012; 150(1-3): 200-7.
[http://dx.doi.org/10.1007/s12011-012-9506-6] [PMID: 22983774]

[127] Jung JY, Kim WJ. Involvement of mitochondrial- and Fas-mediated dual mechanism in CoCl2-induced apoptosis of rat PC12 cells. Neurosci Lett 2004; 371(2-3): 85-90.
[http://dx.doi.org/10.1016/j.neulet.2004.06.069] [PMID: 15519734]

[128] Wiegand H, Uhlig S, Gotzsch U, Lohmann H. The action of cobalt, cadmium and thallium on presynaptic currents in mouse motor nerve endings. Neurotoxicol Teratol 1990; 12(4): 313-8.
[http://dx.doi.org/10.1016/0892-0362(90)90049-I] [PMID: 2168016]

[129] De Boeck M, Kirsch-Volders M, Lison D. Cobalt and antimony: genotoxicity and carcinogenicity. Mutat Res 2003; 533(1-2): 135-52.
[http://dx.doi.org/10.1016/j.mrfmmm.2003.07.012] [PMID: 14643417]

[130] Keen CL, Clegg MS, Hanna LA, *et al.* The plausibility of micronutrient deficiencies being a significant contributing factor to the occurrence of pregnancy complications. J Nutr 2003; 133(5) (Suppl. 2): 1597S-605S.
[http://dx.doi.org/10.1093/jn/133.5.1597S] [PMID: 12730474]

[131] Araya M, Koletzko B, Uauy R. Copper deficiency and excess in infancy: developing a research agenda. J Pediatr Gastroenterol Nutr 2003; 37(4): 422-9.
[http://dx.doi.org/10.1097/00005176-200310000-00005] [PMID: 14508211]

[132] Uriu-Adams JY, Scherr RE, Lanoue L, Keen CL. Influence of copper on early development: prenatal and postnatal considerations. Biofactors 2010; 36(2): 136-52.
[http://dx.doi.org/10.1002/biof.85] [PMID: 20232410]

[133] Toaff R, Toaff ME, Peyser MR, Streifler M. Hepatolenticular degeneration (Wilson's disease) and pregnancy. A review and report of a case. Obstet Gynecol Surv 1977; 32(8): 497-507.

[http://dx.doi.org/10.1097/00006254-197708000-00001] [PMID: 331163]

[134] Oga M, Matsui N, Anai T, Yoshimatsu J, Inoue I, Miyakawa I. Copper disposition of the fetus and placenta in a patient with untreated Wilson's disease. Am J Obstet Gynecol 1993; 169(1): 196-8.
       [http://dx.doi.org/10.1016/0002-9378(93)90163-D] [PMID: 8333453]

[135] Bermúdez L, García-Vicent C, López J, Torró MI, Lurbe E. Assessment of ten trace elements in umbilical cord blood and maternal blood: association with birth weight. J Transl Med 2015; 13(1): 291.
       [http://dx.doi.org/10.1186/s12967-015-0654-2] [PMID: 26346609]

[136] Haddad DS, al-Alousi LA, Kantarjian AH. The effect of copper loading on pregnant rats and their offspring. Funct Dev Morphol 1991; 1(3): 17-22.
       [PMID: 1802039]

[137] Ferm VH, Hanlon DP. Toxicity of copper salts in hamster embryonic development. Biol Reprod 1974; 11(1): 97-101.
       [http://dx.doi.org/10.1095/biolreprod11.1.97] [PMID: 4457127]

[138] Murawski M, Bydłoń G, Sawicka-Kapusta K, et al. The effect of long term exposure to copper on physiological condition and reproduction of sheep. Reprod Biol 2006; 6 (Suppl. 1): 201-6.
       [PMID: 16967100]

[139] Johnson A, Carew E, Sloman KA. The effects of copper on the morphological and functional development of zebrafish embryos. Aquat Toxicol 2007; 84(4): 431-8.
       [http://dx.doi.org/10.1016/j.aquatox.2007.07.003] [PMID: 17714802]

[140] Lewis SS, Keller SJ. Identification of copper-responsive genes in an early life stage of the fathead minnow Pimephales promelas. Ecotoxicology 2009; 18(3): 281-92.
       [http://dx.doi.org/10.1007/s10646-008-0280-3] [PMID: 19020976]

[141] Ganesan S, Anaimalai Thirumurthi N, Raghunath A, Vijayakumar S, Perumal E. Acute and sub-lethal exposure to copper oxide nanoparticles causes oxidative stress and teratogenicity in zebrafish embryos. J Appl Toxicol 2016; 36(4): 554-67.
       [http://dx.doi.org/10.1002/jat.3224] [PMID: 26493272]

[142] Li J, Lock RA, Klaren PH, et al. Kinetics of Cu2+ inhibition of Na+/K(+)-ATPase. Toxicol Lett 1996; 87(1): 31-8.
       [http://dx.doi.org/10.1016/0378-4274(96)03696-X] [PMID: 8701442]

[143] Li J, Quabius ES, Bonga SW, Flik G, Lock RA. Effects of water-borne copper on branchial chloride cells and Na+/K+-ATPase activities in Mozambique tilapia (Oreochromis mossambicus). Aquat Toxicol 1998; 43(1): 1-1.
       [http://dx.doi.org/10.1016/S0166-445X(98)00047-2]

[144] Hoyle I, Shaw BJ, Handy RD. Dietary copper exposure in the African walking catfish, Clarias gariepinus: transient osmoregulatory disturbances and oxidative stress. Aquat Toxicol 2007; 83(1): 62-72.
       [http://dx.doi.org/10.1016/j.aquatox.2007.03.014] [PMID: 17442412]

SUBJECT INDEX

T

Tobacco smoke 156
Toxic effects 10, 13, 38, 67, 81, 87, 105, 108,
      147, 148, 159
  radioactive 159
Toxic heavy metals 98
Tyrosine 70, 71
    residues 71
    Sulfatases 70

V

Vaccine adjuvants 66, 81, 100
Visual hallucination 100
Voltage-dependent calcium channels (VDCC)
      83, 86
Vomiting 3, 4

W

Waste 41, 51, 115, 156
  industrial 156
  mining 41
Water 20, 35, 37, 46, 54, 74, 89, 101, 115,
      148, 150, 156, 163, 166
    contaminated 46, 166
    purification processes 148
    treatment plants 35
Wilson's disease 98, 121, 165
Wnt signaling pathway 73, 161, 162

www.ingramcontent.com/pod-product-compliance
Lightning Source LLC
Chambersburg PA
CBHW041702210326
41598CB00007B/500